Holy Holmium!
Complete General Chemistry in 150 Pages
For High School, University, Business, and Life

Adam Gottlieb

Honours Chemistry graduate of Princeton University
Environmental Toxicology graduate of Concordia University
Chemistry tutor since 1993

Third Edition, 2010
ISBN 978-1-4357-0081-9
Published by lulu.com.

Typeset in Georgia twelve point. Illustrated and laid out by the author. Author photo by Amy Felske, used with permission.

Contents

Check them off as you go!

Introduction

Hello!

I have written this book for you to succeed with chemistry, whether or not:

? You are attending school, doing distance education, or self-teaching;
? You need chemistry for a degree, a job, or another purpose;
? You have ever studied chemistry before;
? You like chemistry so far;
? You've been told you are "good" or "bad" or "slow" or "quick", at it, at math, at science, at school, and so on;
? English is your first language;
? You are a young man or woman or an older one;
? You learn best by seeing, hearing, or doing.

I have been teaching chemistry to all sorts of people, for 17 years. I have developed ways of learning and doing that make sense to almost anyone, and am confident this will include you. I presume that your time and money are as precious as your success.

So many students have asked me, "why don't they make it this clear in school?" This book answers that question. It is meant to be almost as fun and effective as studying with me in person. I'm speaking for myself, of course, because I like tutoring. Still, I wrote this book to reach people like you who may not have handy a qualified and affordable tutor, or who may prefer a printed book you can keep coming back to.

Holy Holmium covers high school and early university chemistry. Each section can be used independently. There's an index so you can refer forward and backward to fill in the blanks. All technical words are defined the first time they are used.

There are only 150 pages, but all of the usual topics are covered in full depth, with examples of solved problems. Some sections may be more than you need. There may also be some special things you need to know that don't appear here. Your teacher or boss can tell you what they are and where to find them, and part of my guarantee is that they will be few enough that you don't need to buy a whole other book. You may find some topics easier and some harder, but by the end you will be able to:

✓ Get it right on the exam;
✓ Use it right on the job;
✓ Understand how it all fits together;
✓ Make sense of chemistry related newspaper and magazine articles, TV and radio shows, movies, and conversations;
✓ Explain chemistry and related science to others; and/or
✓ Find the spirituality and pleasure in them.

It depends on what you need. If you do not gain significant knowledge, understanding, problem-solving skills, or pleasure, I will personally refund your purchase. For this, and feedback in general, there is a form at the end of the book.

What is chemistry, and why am I learning it?

There are certain natural structures of matter. From these, we can learn what matter likes, how to break it down and assemble it in ways not found in nature, and how hard or easy that will be. Historically, this has been to:

✓ Help people;
✓ Understand the sacred;
✓ Make money.

It depends how you look at it. Pretty much everyone agrees that the goal is to make something valuable from cheap, abundant ingredients.

Keep this in mind as we study different branches of chemistry. They are not all developed to the same extent. Some sections might seem incomplete and others too long. It doesn't have to stay that way. Especially the less developed branches are begging for further experiment and theory, and that's part of the scientific method that got us here.

The word "chemistry" comes from "alchemy," which is an English approximation of an Arabic word that in turn comes from Greek, Egyptian, and other sciences. Alchemy is simply the study of how to turn one thing into another. Today it may sound mystical or evoke trying to turn cheap metals into precious gold, which classical chemistry would say is impossible. However, chemistry is basically that: where matter comes from and goes to, and how that takes or gives energy.

Every textbook is a little different.

In terms of sequence, this one is pretty standard: from elements to compounds to their physical and chemical processes. That's what's made the most sense to most of my students. If the sequence is different from your teacher's or text's, rest assured that all of the basic content is here and use the index to find it. For example, you'll see that Lewis structures, VSEPR, and hybridization are covered in the section called "Covalent compounds in detail".

Holy Holmium uses all the standard chemical terms, but its writing style is conversational and to-the-point, and its pages are light and spacious. This is meant to fix common complaints I've heard from students about their textbooks and/or classes being too expensive, big, heavy, time-consuming, intimidating, and/or hard to understand. At the same, my experience is that most people can handle the full technical detail if it's presented right, and that being called "dummies" or taught oversimplifications can be a disservice.

So I hope you're up for the challenge of the task before you – and maybe, just maybe, of finding that chemistry, science, school, work, and/or life is not such a task after all.

Acknowledgments

It takes a lot to bring a person successfully into this world. I have much for which to thank my parents, Richard and Sandra, including a good, broad education and our ongoing exchanges.

At St. George's Elementary and High School, many teachers brought out the best in me and encouraged me on. With apologies to those I don't name, the ones that come to mind are Beatrice Lewis, Serge Rousseau, John Daly, and Michael Hayes, my first chemistry prof. Vice-Principal Bahadur Bhatla helped me get recognition and admission to college.

Once there, Aniko Lysy (Marianopolis College) and Maitland Jones (Princeton University) demanded I understand this stuff deeply and showed me how. Clarence Johnson (now at Spelman College) and Marjorie Carhart continued my pattern of friendship with and mentorship by teachers who let me see the lives that go with the work they love. Peter Bunnell of the Art Museum at Princeton put me in that picture, by giving me a valuable task – the conservation of photographs by Alfred Stieglitz – and Andrew Bocarsly of the Chemistry Department helped turn it into my thesis.

During the Summers, Ken Lum and Claire Lemieux at Environment Canada supervised my first lab job, and it was a good one: ethical work, good pay, and a solid team.

When I finally graduated, the economy was in a recession and there wasn't work for me there or much place else. I started tutoring. I had never though of myself as a social or well-spoken person, but I liked it and built a successful practice. At one point I returned to the lab, and at another I returned to school. Yet I remained a teacher, eventually branching out into popular education and radio broadcasting.

From semester to semester, my energy is renewed by new students. I work mostly with 15- to 22-year-olds. Those ages were difficult for me, and I appreciate in my students both the innate curiosity and ability that still shine as well as the creeping shadow of criticism and pressure. I hope this book, like my tutoring and teaching, can keep your doors open, and even show or build you some new ones.

When a student returns, I feel even more valuable to the community and blessed to know that person, and through them human nature. For this I have to thank, among many, Deja Griffith, Sharon Shlien, Amardeep Khaira, Rupinder Deol, Jackie Haas, Maha Nimeh, Barry Burstein, Lauren Nemes, Sabrina Perri, Evan Perlman, and the Felske, Pagano, Eloi, and Nader families.

Thanks to Wayne Clasper of Marianopolis College Student Services for referring me to many of these fine folks and supervising the quality of my work, which isn't always easy when you run your own business. Likewise to Johanna Mantsinen, who runs the findatutor.ca web site, and to Robert Schweitzer who referred me. I am also grateful to Dr. Dave Berry (University of Victoria) for taking the time to orient me and much more, such that I could wind up teaching labs there.

In the publishing process, Don Campbell at lulu.com was very helpful and inspiring, and so was Orlando Rojas at locutorio.net. Evan Perlman did an amazing job preparing contact lists for the ensuing publicity work.

When I was in high school, I got to watch James Burke's TV series, *The Day the Universe Changed*. It gave me a picture of the connections I felt between all of the sciences, and between science and life, and taught and inspired me much further. Now I realize that countless filmmakers, musicians, writers, athletes, and other public people, like Bob Sinclair, Wyclef Jean, Bruce Springsteen, Ken Dryden, India Arie, Jack Johnson, Starhawk, Matt Groening, J.R.R. Tolkien, Dumisani Maraire & Ephat Mujuru, Ry Cooder & Vishwa Mohan Bhatt, and Daniela Mercury, have also been nurturing my dreams and ability to fulfill them, all along.

Then there are my brothers and friends over the years, who complete a good education and life. Thanks in particular to Devlin Kuyek, Jon Gottlieb, Lucy Tomasetta, Maya Goodrich, Mélissa Guay, Appiah Joseph Kojo Annan, Françoise Giroux, Eddie Fernandez Ureche, Jenni Huntly, Holly Mackay, Bruce Rutley, Francine Charpentier, Rae Shepp, Orlando Rojas, Jordie Robinson, Steven Pratt, Jorden Leighton, and the Thomas-Archambault family for joining me in making days special and the future bright.

Other living role models for being a super and evolving person in this world include, at the very least, Loris Mirella, Michael Cooke, Mark Messier, Andrew Harwood, Nancy Stark-Smith, and Adrian Harewood.

Thank you.

About the Author

Adam Gottlieb is a 38 year-old Canadian who has been tutoring high school and university chemistry since 1993, with teaching assistanceships at Concordia University and the University of Victoria. He has an Honours Chemistry bachelor degree from Princeton University and a graduate diploma in Environmental Toxicology from Concordia University. Over the years he has contributed to community organizations as a popular educator and radio journalist, branching out into nutrition, agriculture, ecology, and human rights. He also does shiatsu massage, contact improvisation dance, hockey, and other body arts.

Reviews from Students and Teachers

"Adam is precise and gets to the point. I caught up fast and was able to understand everything. I got a high enough grade to meet all the requirements for pre-med at college." – Mélanie B.

"He's skilful at making complex ideas seem approachable by breaking things down in steps, introducing different methods of problem solving, and providing real life examples. Thanks to Adam, not only did I pass my course, but I now have gained respect for Chemistry as well as a fundamental understanding of the subject that will empower me later in my studies. He clearly has an extensive amount of experience and qualifications with kids, teens, and adults." – Marcela L.

"I acquired a certain confidence that I had lost because of poor marks." – Marie-Anne P.

"It helped me realize that any course can be comprehensible if taught in a way that the student can understand. As a result, I am less likely to give up easily when I find a course very difficult. Adam's caring and sense of humour make the school experience far less stressful and more enjoyable. He makes sure the student has everything he or she needs to succeed. It is obvious that he takes his job seriously." – Angela I.

"I could always rely on Adam to explain what I had trouble with. I had peace of mind." – Raluca P.

"When I was referred to Mr. Gottlieb, he was the busiest tutor as well as being the most recommended. He always kept his word about coming on time and coming well prepared. I enjoyed his friendly approach and did very well in the course. He does not see students' questions as stupid; on the contrary, he answers them with great flexibility." – Liliane R.

"I am now more capable of asking and answering questions, thus a more active participant in class. Most importantly, I feel more confident about starting organic chemistry next semester." – Sabrina P.

"My marks went from 66% after the first midterm to 83% on the course! How do you think I feel? I'm thrilled." – Robert R.

"It helped me clear up some misconceptions & gain a better understanding of basic concepts in the field. In addition, I achieved high enough marks in my course that I was accepted to the university program of my choice. It was helpful and pleasant." – Eric N.

"Adam's chemistry notes really assisted me in my studies, and gave me an edge. A lot of what I understood in Chemistry 101 was a direct result of his notes. They are exceptional, thorough, and concise." – Deja G.

"Adam taught me everything I needed to know for the final exam. He was much better than previous chemistry tutors I've had." – Matthew W.

"My exam went incredibly well. I enjoy chemistry now. I understand what I'm doing and why." – Anne W.

"I was able to finish my assignments (which were extremely long) on time for the quiz. I had so much more confidence during class. Since my problems decreased in this course I was able to concentrate myself more on my other courses. Getting a good grade in chemistry helped me also to get into the field I wanted in university. I wasn't comfortable with the idea of getting help since I never had to do so before, but Adam made it easy for me. He's very understanding and caring. He's conscious that each person has a different way of understanding things." – Maha N.

"I gained the confidence I needed to pass the course, something I had come to believe was impossible. The little jokes he throws in, especially with difficult concepts, always cheered me up. His notes are clear and concise, focusing on key concepts followed by examples." – Jane B.

"He definitely boosted my confidence in the domain which I was studying and helped me meet all my immediate needs." – Robert A.

"A source of relief when on the verge of extreme frustration and despair with class." – Tara M.

"It was an excellent learning experience that helped me improve my study habits." – Matthew K.

"I found him to be outstanding. Adam is very knowledgeable on the subject of Chemistry and has the ability to explain concepts in a clear and simple manner that make it easy to apply to problems. I found him to be very encouraging and he seemed to really care about me getting a better understanding of chemistry." – Catherine P.

"He was exceptionally qualified to help me. From the first lesson, a specific topic that I had struggled in for weeks got cleared up in less than half an hour. His notes were irreplaceable, as they were precise and contained comprehensive charts, instructions or examples, yet were always produced naturally. Adam's efficiency and experience are such that he can teach an entire semester's course in only eight to ten hours. His confidence in his work gave me the confidence to practice and study outside class." – Evan P.

"Adam's patient and caring nature seems to lend themselves to effective teaching. His knowledge of the subject matter and effective communications skills make him an ideal educator. He can explain even the most complex topics." – Barry B.

"As my teaching assistant, Adam treated students with respect and always offered assistance that was supportive and constructive. He is organized and has initiative and ability." – Professor Geoff Selig

"Adam was invited to deliver a lecture to my class. He is a compelling speaker and encourages people to participate. He is sincere and committed to the work he does." – Professor Yasmin Jiwani

Atoms:
The Building Blocks

Theories of the atom

Many cultures have had the idea that there must be some irreducible substance things are made of. Two hundred years ago in England, John Dalton called this the atom. He assumed it had to be round, perhaps because this is a stable form for many free objects shaped by natural forces, from big planets to little grains of sand.

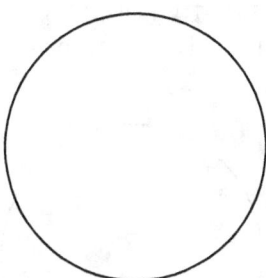

Alchemists already understood that atoms of one substance, like gold, are different from atoms of another, like iron, and they also knew how to tell the difference between one of these pure substances and a mixture, like brass. What Dalton and his followers did was give us a visual model for how this happens on a microscopic level: atoms bond together to make molecules. They can combine for different reasons and in different proportions, as we'll see in the next chapter. A collection of one kind of atom is called an element. A collection of one kind of molecule is called a compound.

In general, chemists consider that there are only about a hundred elements out of which everything in the world is made, and relatively few of these are naturally abundant.

- The earth is 49% oxygen (O), 26% silicon (Si), 8% aluminium (Al), 5% iron (Fe), 3% calcium (Ca), and less than 10% everything else combined.
- Water bodies are 89% O and 11% hydrogen (H), with traces of some other elements.
- The air is 78% nitrogen (N), 21% O, 1% argon (Ar), with bits of a few others.
- Our bodies are 65% O, 18% carbon (C), 10% H, 3% N, 2% Ca, 1% phosphorus (P), and a little of some other elements.

Eventually, people came along and challenged the idea that the atom has no parts. They relied on new theories of electricity, magnetism, and radioactivity, and equipment with which to measure them.

Joseph John Thomson found that atoms attract and repel negatively charged objects, so must have positively and negatively charged particles, called protons (p^+) and electrons (e^-).

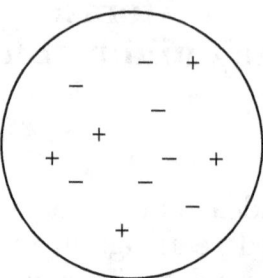

Ernest Rutherford, Hans Geiger, and Ernest Marsden said fine, but the protons aren't just anywhere. If we throw positive objects at an atom, most of the time they're attracted straight through, and only in the centre (nucleus) do they get repelled, so that must be where the protons reside.

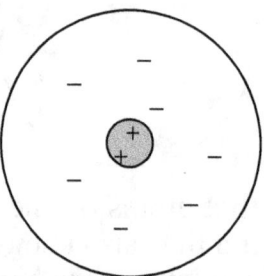

James Chadwick said true enough, but what's this uncharged particle being ejected from the centre of some radioactive substances? Let's call it a neutron (n^o). Together, p^+ and n^o are called nucleons.

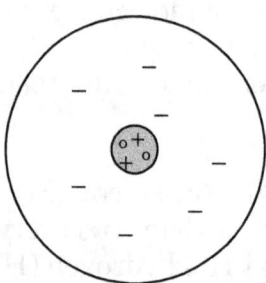

All of this was still within Dalton's classical mindset of hard and fast objects. Quantum science came along and picked that apart.

It said, each of the three kinds of atomic particles is actually made up of a combination of subatomic particles (mesons, bosons, neutrinos, and so on), and this combination is ever changing! In other words, matter is a wave-like flow of energy. That's what Albert Einstein's famous "e (energy) = m (mass) c (speed of light) squared" equation says.

And it said the electrons aren't just anywhere within the space of the atom, there's a logic to them.

People had been experimenting with elements, by doing things like heating them and then watching them emit the energy back as light. They noticed that each element gave off only certain colours of light, and this was different from one element to the next. For example, hydrogen emits mostly in the blue-violet part of the visible spectrum:

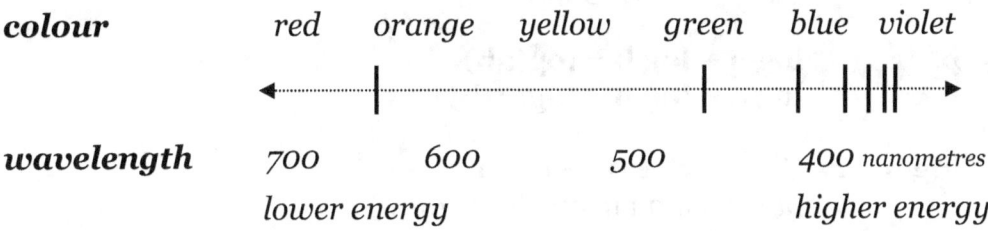

Niels Bohr explained why. Electrons and protons attract each other because they have opposite charge. Protons are much heavier than electrons. So when an electron moves toward the protons, it becomes more stable and releases energy which often happens to be in or near the visible light range of the electromagnetic spectrum. If we only see certain exact colours, it is because electrons can only transition between certain fixed distances from the protons. Electrons exist in shells, much like planets orbit a sun. Except that you can have more than one electron in the same orbit.

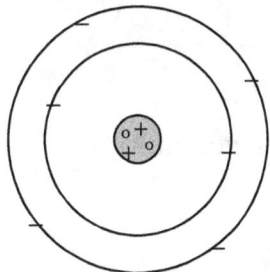

To situate things, here's the entire electromagnetic spectrum of wave-particle energy:

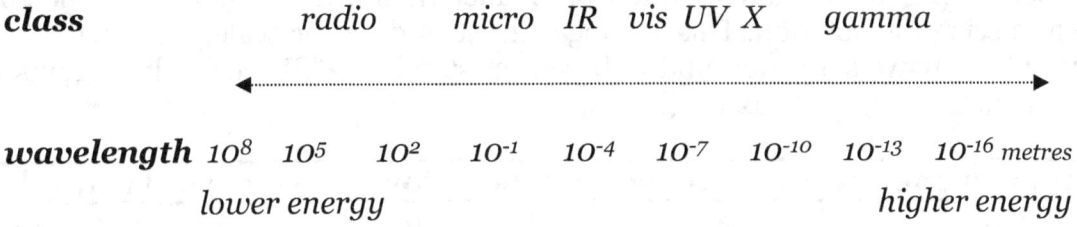

This scale uses a logarithmic scale, which the next section explains. "IR" stands for infra-red, "vis" for visible, "UV" for ultra-violet, and "X" for X-rays.

A word about math

Math comes up a lot in general chemistry. Often, a student of mine will think they don't understand the chemistry, when really it's the math that's getting in the way. Now is a good time to learn things you never did and clear up old confusions.

The basic algebraic shortcuts you will need in this course are:

Given $a = b/c$ Given $a = b^c$
therefore $b = ac$ $\log a = \log(b^c)$
and $c = b/a$ $\log a = (c)(\log b)$

Given $a = b^c$ $\log a + \log b = \log(ab)$
$b = \sqrt[c]{a}$ $\log a - \log b = \log(a/b)$

Given $a = \log b$ Given $ax^2 \pm bx \pm c = 0$
$b = 10^a$ (an equation in quadratic form)
                                          $x = (-b \pm \sqrt{(b^2-4ac)})/2a$

Given $a = \ln b$
$b = e^a$

$\pm$ means "plus or minus" and $\approx$ means "roughly equal to".
To do calculations with percentages, convert them to decimals by dividing by 100. For example:

   $47\% = 47/100 = 0.47$
   $8\% = 8/100 = 0.08$
   $0.93\% = 0.93/100 = 0.0093$

To convert a decimal to a percentage, multiply by 100. For example:

   $(0.47)(100) = 47\%$
   $(0.08)(100) = 8\%$
   $(0.0093)(100) = 0.93\%$

Numbers bigger than a few hundred or smaller than a few hundredths tend to be written in scientific notation. This is a logarithmic and metric scale, meaning it works in multiples ("powers") of ten and is based on standard ("SI") units like grams (g), litres (L), metres (m), and Pascals (Pa).

Common name	Scientific name	Symbol	Power	Numerical equivalent
trillion	tera	T	10^{+12}	1,000,000,000,000
billion	giga	G	10^{+9}	1,000,000,000
million	mega	M	10^{+6}	1,000,000
thousand	kilo	K	10^{+3}	1,000
hundred	hecto	H	10^{+2}	100
ten	deca	D	10^{+1}	10
unit	-	-	10^{0}	1
tenth	deci	D	10^{-1}	0.1 *or* 1/10
hundredth	centi	C	10^{-2}	0.01 *or* 1/100
thousandth	milli	M	10^{-3}	0.001 *or* 1/1,000

millionth	micro	μ	10^{-6}	0.000 001
				or 1/1,000,000
billionth	nano	N	10^{-9}	0.000 000 001
				or 1/1,000,000,000
trillionth	pico	P	10^{-12}	0.000 000 000 001
				or 1/1,000,000,000,000

So, for example:

5403 could be written as 5.403×1000 or 5.403×10^3
(A calculator or computer may show it as 5.403E3, 5.403EE3, 5.403EXP3, and so on.)
0.069 is 6.9×10^{-2}
0.000 000 5 is 5×10^{-7}
89,000,000 is 8.9×10^7

A number is said to have a certain number of significant figures ("sig figs"), according to the following rules:

○ Nonzero digits are significant;
○ Zeroes trapped between them are significant;
○ Zeroes between them and the decimal point are not significant.

For example:

5403 has 4 sig figs
0.069 has 2
0.000 000 5 has 1
89,000,000 has 2
2.00 has 3

The easiest way to see this is to put it in scientific notation. For example:

5.403×10^{-3} has 4
6.9×10^{-2} has 2
5×10^{-7} has 1
8.9×10^7 has 2

When asked to solve a numerical problem, notice how many sig figs are in each piece of data given. In general, whoever has the fewest, that's how many sig figs to use in your answer. But, to prevent rounding error: do calculations in 1 more sig fig than this; for each step, reuse the exact result the calculator gave you in the last step, rather than retyping it in approximately.

Technically, that is the convention for multiplying and dividing. For adding and subtracting, you'd normally report the answer to the last significant decimal place shared by the inputs. But since most problem solving involves a mix of adding, subtracting, multiplying, and dividing, the rules get blurred and the convention is to follow the multiplying/dividing rule.

Emission spectra

The strength of p^+ attraction that holds an e^- in its orbit can be calculated:

$E = -R^*Z^2(1/n^2)$
E is energy, whose units are J, named for James Joule
R^* is a constant, 2.178×10^{-18} J
Z is the number of protons (atomic number) characteristic of that element
n is the shell number

J are equal to $kg \cdot m^2/s^2$. Another unit used for energy is the electron-volt (eV), equal to 1.61×10^{-19} J. The minus sign means the e^- is more stable being there than being free. So, this tells you how much energy you'd have to put in to remove the electron.
By extension, the energy of an e^- jump between two shells is:

$E = R^*Z^2(1/n^2_{final} - 1/n^2_{initial})$

Different values of n_{final} give different kinds of energy, and have special names.

n_{final}	Kind of light	Name
1	Ultraviolet	Lyman
2	Visible	Balmer
3	Infrared	Paschen
4	Infrared	Brackett

These equations consider proton-electron forces in isolation. So they only work perfectly for the first element, hydrogen (H), which has one proton and one electron. The moment you have multiple protons or electrons, you have forces among them, and inner electrons partly block protons' effect on outer electrons. You can calculate what fraction ($Z_{effective}$) of the total proton pull an electron feels in practice, using rules developed by John Slater, but that's beyond the scope of basic chemistry. Look it up in a book or on the internet, if you like – but, to understand it, you first need to know the quantum numbers and a couple more equations.

$E = h\nu$
h is Max Planck's constant, 6.626×10^{-34} J$\cdot$s
ν ("nu") is the frequency, how many wave crests pass in one second (waves/s, 1/s, Hertz, Hz)

$v = c/\lambda$
c is the speed (velocity) of light, 2.998×10^8 metres per second (m/s)
λ ("lambda") is the wavelength, in m

$\lambda = h/mv$
m is the mass of the electron, 9.109×10^{-31} kilograms (kg)
v is its speed, in m/s
This is called Louis de Broglie's equation.

Another way of showing the energy of e⁻ transition between shells is:

$1/\lambda = RZ^2(1/n^2_{initial} - 1/n^2_{final})$
$1/\lambda$ is also called wavenumber ($\bar{v}$, "nu bar"), in 1/m
R is Johanes Rydberg's constant, 1.097×10^7 1/m

Compared to the previous equation for e⁻ jumps, here $n_{initial}$ and n_{final} have switched places, to give a positive answer, because wavelength has to be positive.

> *Practice problem:*
> *What colour of light is emitted by the third Balmer line in hydrogen?*
>
> Z = 1 for hydrogen
> n_{final} = 2, since it's a Balmer line
> $n_{initial}$ = 5: the 1st Balmer line is 3→2, the 2nd is 4→2, the 3rd is 5→2
>
> The second equation for e⁻ jumps is handier here.
> $1/\lambda = RZ^2(1/n^2_{initial} - 1/n^2_{final})$
> $1/\lambda = (1.097 \times 10^7 \text{ 1/m})(1^2)(1/2^2 - 1/5^2) = 2.304 \times 10^6$ 1/m
> $\lambda = 1/(2.304 \times 10^6 \text{ 1/m}) = 4.341 \times 10^{-7}$ m
> This is blue light.
> Notice that the answer is in 4 sig figs, which fits the data given. (2 and 5 don't count as data, as they are counting numbers.)

Quantum numbers

Once we've figured out how a single electron behaves in an atom, we can look at how electrons behave collectively. The first step is to give each one a unique label, so we know who we're talking about.

n is the **principal** quantum number, also known as the **shell**
We refer to the 1st, 2nd, 3rd, etc. shells as n = 1, 2, 3, ... up to infinity (∞).

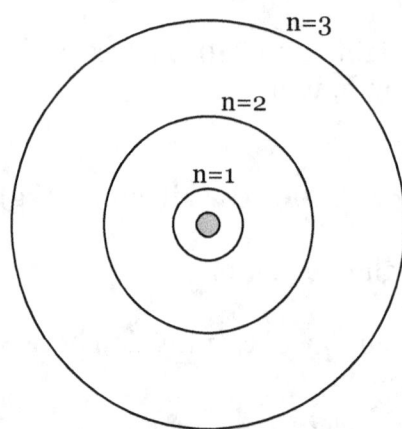

Why infinity? Normally, the Aufbau ("buildup", in German) rule says, electrons go to the lowest shell that has free space. We'll see in a second how you know how many electrons each shell can take. If you bombard an atom with energy, electrons jump to shells further from the nucleus. When you stop supplying energy, the electrons come back in, and emit energy you can measure. In other words, the balls we usually draw for atoms have fuzzy edges: there's always a slim chance you could find one of its electrons further away.

l is the **azimuthal** quantum number or **subshell**
By convention, for a given n, l can = 0, 1, 2, 3, ... up to (n-1).
Also by convention, 0 is called "s", 1 is "p", 2 is "d", 3 is "f", 4 is "g", and so on.

Only s, p, d, and f are needed for natural elements. g and subshells beyond could be asked in a purely theoretical question. So:

If n = 1, l can = 0: we have only an s subshell
If n = 2, l can = 0, 1: we have an s and a p subshell
If n = 3, l can = 0, 1, 2: s, p, d
and so on.

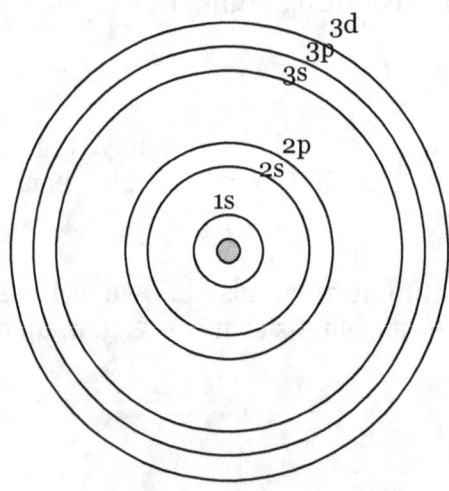

m_l is the **magnetic** quantum number or **orbital**
For a given l, m_l can = -l up to +l. So:

If l = 0, m_l can = 0: we have 1 possible value, 1 orbital
If l = 1, m_l can = -1, 0, +1: 3 possible values, 3 orbitals
If l = 2, m_l can = -2, -1, 0, +1, +2: 5 orbitals
and so on.

Each subshell, like the atom, has to be round. So how do we cut it up into parts? Given the three-dimensional coordination system:

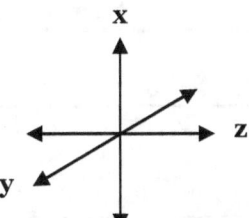

The orbitals are:

l = 0: 1 orbital

s

l = 1: 3 orbitals

p_x p_y p_z

l = 2: 5 orbitals

d_{xy} d_{xz} d_{yz} d_{x2-y2} d_{z2}

...   ...

So, for example,

p_x + p_y + p_z =

Likewise, the d's add up to a sphere, and so do the f's.

Holy Holmium! Complete General Chemistry in 150 Pages

Shaded and unshaded regions represent spaces with opposite electrical charge. One is negative, the other is positive. It doesn't matter which is which. And the electrons can be found in either one.

m_s is the **spin** quantum number or **electron**
For any orbital, by convention, m_s can $= -\frac{1}{2}$ or $+\frac{1}{2}$

This says, any orbital can have up to two electrons, and they must be orbiting in opposite directions. This is the Wolfgang Pauli exclusion principle. So:

n	l	m_l	# orbitals	Maximum # e⁻
1	0 (s)	0	1	2
				Total: 2
2	0 (s)	0	1	2
	1 (p)	-1, 0, +1	3	6
				Total: 8
3	0 (s)	0	1	2
	1 (p)	-1, 0, +1	3	6
	2 (d)	-2, -1, 0, +1, +2	5	10
				Total: 18
...	...	...	...	...

This is why, in the periodic table, we'll see that the s block is 2 elements wide, p is 6, d is 10, and f is 14.

Which of the following sets of quantum numbers are allowed for an electron?
a) n 4, l 3, m_l -2, m_s 0
b) n 4, l 4, m_l 0, m_s -½
c) n 4, l 2, m_l -1, m_s +½
d) n 4, l 2, m_l 3, m_s -½

a) no, because m_s can't be 0
b) no, because l can only go up to n-1, which here is 3
c) yes
d) no, because m_l can only go from −l to +l

Here's where drawing the shells, subshells, and orbitals starts to get complicated. Box-orbital diagrams provide a convenient shorthand. They show orbitals as boxes and electrons as arrows. For example, the first e⁻ in an l = 0 (s) subshell is:

↑

And the second e⁻ is:

↑↓

Notice that they have opposite spin.

If we go to a subshell with many orbitals, Friedrich Hund's rule says it's most stable to keep electrons in separate orbitals, with parallel spin, for as long as you can. For example, in an l = 1 (p) subshell:

1st e⁻: ↑ __ __

2nd e⁻: ↑ ↑ __ (or ↑ __ ↑ , or __ ↓ ↓ , and so on)

3rde⁻: ↑ ↑ ↑

4th e⁻: ↑↓ ↑ ↑

5th e⁻: ↑↓ ↑↓ ↑

6th e⁻: ↑↓ ↑↓ ↑↓

A set of quantum numbers which doesn't respect Pauli's rule is "forbidden". One that disrespects Hund's or Aufbau's is "excited". The one that respects all three is "ground".

Which of the following sets of quantum numbers are ground states, which are excited, and which are forbidden, for two electrons in a d subshell? Show box-orbital diagrams.

a) 1st e⁻: n 5, l 2, m_l -1, m_s +½
   2nd e⁻: n 5, l 2, m_l -2, m_s +½

b) 1st e⁻: n 5, l 2, m_l -1, m_s +½
   2nd e⁻: n 5, l 2, m_l -1, m_s -½

c) 1st e⁻: n 5, l 2, m_l -1, m_s +½
   2nd e⁻: n 5, l 2, m_l -1, m_s +½

d) 1st e⁻: n 5, l 2, m_l -1, m_s +½
   2nd e⁻: n 5, l 2, m_l 0, m_s -½

a) ↑ ↑ __ __ __
   Ground

b) ↑↓ __ __ __ __
   Excited: Hund says, put them in separate orbitals

c) ↑↑ __ __ __ __
   Forbidden

d) ↑ ↓ __ __ __
   Excited: Hund says, make them parallel

Notice that it doesn't matter whether you call the first of these orbitals m_l -2, m_l -1, and so on, as long as the others are different. Likewise, an arrow pointing up could be +½ or -½; but if you decide that +½ is "down", for example, then -½ has to be "up".

Configurations in which all e- are paired are diamagnetic, which means that they are repelled by an outside magnetic field. The moment you have unpaired e-, the element is paramagnetic and attracted to an outside field. The more unpaired e- (net spin) it has, the stronger the attraction.

Electronic configuration and the periodic table

This is a way of summarizing all of the electrons in an atom, in one shot.

1st element, H, 1e- (and 1p+), configuration $1s^1$
2nd element, He, 2e- (and 2p+), configuration $1s^2$
3rd element, Li, 3e-, $1s^2 2s^1$

If you're confused at this point, look back at the notes for the l quantum number.

4th element, Be, $1s^2 2s^2$
5th, B, $1s^2 2s^2 2p^1$
6th, C, $1s^2 2s^2 2p^2$
7th, N, $1s^2 2s^2 2p^3$
10th, Ne, $1s^2 2s^2 2p^6$
11th, Na, $1s^2 2s^2 2p^6 3s^1$
and so on.

Notice that if you add all the superscripts in an element's e- configuration, you get its total number of e-.

All of the elements with the same ending (valence) are put in the same column in the periodic table, which organizes elements in order of atomic number.

1 H																	2 He
3 Li	4 Be											5 B	6 C	7 N	8 O	9 F	10 Ne
11 Na	12 Mg											13 Al	14 Si	15 P	16 S	17 Cl	18 Ar
19 K	20 Ca	21 Sc	22 Ti	23 V	24 Cr	25 Mn	26 Fe	27 Co	28 Ni	29 Cu	30 Zn	31 Ga	32 Ge	33 As	34 Se	35 Br	36 Kr
37 Rb	38 Sr	39 Y	40 Zr	41 Nb	42 Mo	43 Tc	44 Ru	45 Rh	46 Pd	47 Ag	48 Cd	49 In	50 Sn	51 Sb	52 Te	53 I	54 Xe
55 Cs	56 Ba	57 La	72 Hf	73 Ta	74 W	75 Re	76 Os	77 Ir	78 Pt	79 Au	80 Hg	81 Tl	82 Pb	83 Bi	84 Po	85 At	86 Rn
87 Fr	88 Ra	89 Ac	104 Rf	105 Db	106 Sg	107 Bh	108 Hs	109 Mt	110 Ds	111 Rg	112	113	114	115	116	117	118

58 Ce	59 Pr	60 Nd	61 Pm	62 Sm	63 Eu	64 Gd	65 Tb	66 Dy	67 Ho	68 Er	69 Tm	70 Yb	71 Lu
90 Th	91 Pa	92 U	93 Np	94 Pu	95 Am	96 Cm	97 Bk	98 Cf	99 Es	100 Fm	101 Md	102 No	103 Lr

s block   p block   □ d block   f block

Some sections have special names:

Section	Name
s and p	Main group elements
d	Transition elements
1st row of f	Lanthanoids
2nd row of f	Actinoids
1st column of s	Alkali metals
2nd column of s	Alkaline earth metals
4th column of p	Chalcogens
5th column of p	Halogens
6th column of p	Noble gases

For ease of printing, and because the f block elements don't get used much in general chemistry, the periodic table is usually printed this way:

1 H																	2 He
3 Li	4 Be											5 B	6 C	7 N	8 O	9 F	10 Ne
11 Na	12 Mg											13 Al	14 Si	15 P	16 S	17 Cl	18 Ar
19 K	20 Ca	21 Sc	22 Ti	23 V	24 Cr	25 Mn	26 Fe	27 Co	28 Ni	29 Cu	30 Zn	31 Ga	32 Ge	33 As	34 Se	35 Br	36 Kr
37 Rb	38 Sr	39 Y	40 Zr	41 Nb	42 Mo	43 Tc	44 Ru	45 Rh	46 Pd	47 Ag	48 Cd	49 In	50 Sn	51 Sb	52 Te	53 I	54 Xe
55 Cs	56 Ba	71 Lu	72 Hf	73 Ta	74 W	75 Re	76 Os	77 Ir	78 Pt	79 Au	80 Hg	81 Tl	82 Pb	83 Bi	84 Po	85 At	86 Rn
87 Fr	88 Ra	103 Lr	104 Rf	105 Db	106 Sg	107 Bh	108 Hs	109 Mt	110 Ds	111 Rg	112	113	114	115	116	117	118

57 La	58 Ce	59 Pr	60 Nd	61 Pm	62 Sm	63 Eu	64 Gd	65 Tb	66 Dy	67 Ho	68 Er	69 Tm	70 Yb
89 Ac	90 Th	91 Pa	92 U	93 Np	94 Pu	95 Am	96 Cm	97 Bk	98 Cf	99 Es	100 Fm	101 Md	102 No

The valences – with a couple of exceptions pointed out below – are:

s^1	s^2	d^1	d^2	d^3	d^4	d^5	d^6	d^7	d^8	d^9	d^{10}	p^1	p^2	p^3	p^4	p^5	p^6

f^1	f^2	f^3	f^4	f^5	f^6	f^7	f^8	f^9	f^{10}	f^{11}	f^{12}	f^{13}	f^{14}

The elements were organized this way, by Dmitri Mendeleev, many years before the quantum system was developed. How did he know to do this?

It turns out that elements with similar valence tend to behave similarly. Perhaps the most striking example is, all of the elements with a p^6 valence are gases and are so stable they exist in nature as individual atoms, almost never combining with other atoms to make molecules.

Meanwhile, all the other main group elements like to gain or lose electrons to acquire a p^6 configuration. In the process, they become charged particles (ions). When a neutral atom gains an e^- (is reduced), it becomes negatively charged (anionic). When it loses (is oxidized), it becomes positive (cationic).

For an s^1 element, for example, the fastest way to become like a noble gas is to lose an e^-, so it will tend to make a +1 ion. Meanwhile, a p^5 likes to gain an e^- to become a -1 ion. Some elements can lose and gain. Here are common charge (oxidation) states of the main group elements:

Column number	I	II	III	IV	V	VI	VII	VIII
Noble-gas-like oxidation states	+1 0	+2 0	+3 0	+4 0 -4	+5 0 -3	+6 0 -2	+7 0 -1	0
Valence	s^1	s^2	p^1	p^2	p^3	p^4	p^5	p^6

Some elements can also lose a few e^- without going all the way down to p^6. For example, C can lose $2e^-$ to become s^2 or $4e^-$ to become p^6. Because of its unique position, H can also gain an e^- to acquire the same configuration as the noble gas He. Meanwhile, the d and f block elements can have many different oxidation states, usually positive. For example, Cr can exist as +2, +3, or +6.

To get the e^- configuration of a + ion, remove the outermost e^-. For a – ion, add an e^-. For example:

C is $1s^22s^22p^2$

C^+ is $1s^22s^22p^1$
C^{+2} is $1s^22s^2$
C^{+3} is $1s^22s^1$
and so on.

C^- is $1s^22s^22p^3$
C^{-2} is $1s^22s^22p^4$
C^{-3} is $1s^22s^22p^5$
C^{-4} is $1s^22s^22p^6$
C^{-5} is $1s^22s^22p^63s^1$
and so on.

For transition elements only, a convention says, when making + ions, remove s e^- before d ones. So, for example:

Ti is $...4s^23d^2$

You'd think Ti^+ is $...4s^23d^1$
But really it's $...4s^13d^2$

Ti^{+2} is $...3d^2$
Ti^{+3} is $...3d^1$

and so on.

Why? It happens that the s and d subshells are very close in energy. For the left-most columns in the d-block, the d is further from the nucleus, so that is where an e⁻ would be removed from. For the right-most columns, the d has gotten so filled with e⁻ that it got sucked in closer to the nucleus, leaving s outermost. To simplify, we assume s is always on the outside. But that may actually be more confusing!

Now, after all that work on electrons, what about protons and neutrons?

To be neutral, the number of e⁻ in an element must be equal to the number of p⁺. However, the number of neutrons can change. For example, most of the time in nature, H has 1e⁻, 1p⁺, and 0n⁰. Occasionally it has 1e⁻, 1p⁺, and 1n⁰, and very occasionally it has 1e⁻, 1p⁺, and 2n⁰. These different forms of a same element are called isotopes.

The symbol for an isotope of an element is:

$$\overset{\text{Mass \#}\atop(p^+ + n^o)}{\underset{\text{Atomic \#}\atop(p^+)}{\text{X}}}\overset{\text{Charge}}{{}^{(p^+ - e^-)}}$$

p⁺ and n⁰ weigh about two thousand times more than e⁻. So, we say that the mass number of the isotope is roughly equal to the #p⁺ + #n⁰. Meanwhile, the mass of the element is an average of its isotopes' masses, in proportion to their natural abundance. So, for example, if H exists 99.985% of the time as ¹H and 0.015% of the time as ²H:

$$(99.985\%)(1) + (0.015\%)(2) = (0.99985)(1) + (0.00015)(2) = 1.00015$$

A final nuance is that, in practice, the mass of an isotope is not measured simply by counting the number of p⁺ and n⁰. Rather, it is calculated by reference to an experimentally determined standard, which happens to be the mass of ¹²C: 1 atomic mass unit (amu) is equal to $1/12$ of the mass of ¹²C. Moreover, it turns out to be a little less than the sum of p⁺ and n⁰, for reasons we'll see in a moment.

So, for example, given that the mass of ¹H is 1.0078 amu and ²H is 2.0140 amu, the weighted average is:

$$(99.985\%)(1.0078 \text{ amu}) + (0.015\%)(2.0140 \text{ amu}) = 1.008 \text{ amu}$$

This is the mass the periodic table gives for that element.

Fill in the missing information for the following isotopes:

Symbol	Atomic #	Mass #	Charge	p⁺	n⁰	e⁻
$^{19}_{9}\text{F}^-$						
				26	30	24
		243	0		148	

a) 9 is the atomic number and #p+
19 is the mass, equal to p+ + n°, so there are 10 n°
-1 is the charge: if the element were neutral, #e- = #p+; here we have 1 extra e-, so there are 10e- total

b) 26 p+ means atomic number 26, element Fe
p+ + n° gives mass, 56
Neutral Fe would have 26e-, here we're missing 2, so the charge is +2
The symbol is $_{26}^{56}\text{Fe}^{+2}$

c) Mass 243 is p+ + n°, so p+ = 243 − 148 = 95, atomic number 95, element Am
0 charge means 95 e- as well
The symbol is $_{95}^{243}\text{Am}$

Moles

In practice, amu's are much smaller than what you can easily measure or ask someone for in real life. What's usually needed is a certain weight (mass) or volume of the substance. Amadeo Avogadro figured out the relationship.

1 mole (mol) of an element contains 6.022×10^{23} atoms
1 mol of a compound contains 6.022×10^{23} molecules
$1\text{ g} = 6.022 \times 10^{23}$ amu

So, for example:

1 atom of H weighs 1.01 amu
1 mol of H = 6.022×10^{23} atoms of H = 1.01 g (molar mass of H)

1 atom O = 16.00 amu
1 mol O = 6.022×10^{23} atoms of O = 16.00 g

1 molecule of H_2O weighs (2)(1.01 amu) + (1)(16.00 amu) = 18.02 amu
1 mol H_2O = 6.022×10^{23} molecules of H_2O
= (2)(6.022×10^{23}) atoms H + (1)(6.022×10^{23}) atoms O = 1.8×10^{24} atoms
or (2)(1.01 g) + (1)(16.00 g) = 18.02 g

In other words, you get elements' molar masses from the periodic table, and add them to get molecules' molar masses.

What is the molar mass, in g, of:
a) P
b) P_2O_5

a) P is 30.97
b) P_2O_5 is $2(30.97) + 5(16.00) = 141.94$

Moles are the fundamental amount of substance in chemistry. They aren't measured directly, but are calculated from whatever you measure.

$n = m/M_M$
n is moles
m is mass, in g
M_M is molar mass, in g/mol

You can get mass from volume, if you know the density:

$m = \rho V$
ρ is density, in g/mL
V is volume, in millilitres (mL)

The reference point for density (1 g/mL) is water at 4 °C (degrees Anders Celsius) or 39 °F (degrees Daniel Fahrenheit).
For gases only:

$n = PV/RT$
P is pressure, in atmospheres (atm), kilopascals (kPa), millimetres of mercury (mmHg), or Evangelista Torricelli units (Torr)
V is in L
R is the gas constant, 0.0821 L·atm/mol·K, or 8.31 L·kPa/mol·K, or 62.4 L·mmHg/mol·K, or 62.4 L·Torr/mol·K
T is temperature, in K, named for Lord Kelvin

It can be confusing to have so many different unit systems for each measurement! Here's how to convert between them:

Property	Conversion factors
Volume	1 L = 1000 mL = 1000 cm³ = 1000 cc
Pressure	1 atm = 101.3 kPa = 760 mmHg = 760 Torr
Temperature	K = °C + 273.15
	°F = (1.8)(°C) + 32

How much do 2.5×10^7 atoms of sulfur weigh?

$n = 2.5 \times 10^7$ atoms$/6.022 \times 10^{23}$ atoms/mol $= 4.15 \times 10^{-17}$ mol
$(4.15 \times 10^{-17}$ mol$)(32.06$ g/mol$) = 1.3 \times 10^{-15}$ g

How many atoms of O are there in a 5.0 L sample of SO_3 at 30 °C and 95 kPa?

T in K = 30 °C + 273 = 303
n = PV/RT = (95 kPa)(5 L)/(8.31 L·kPa/mol·K)(303 K) = 0.189 mol
0.189 mol SO_3 contain (3)(0.189) mol O = 0.566 mol O
(0.566 mol O)(6.022x10²³ atoms/mol) = 3.4x10²³ atoms O

Nuclear reactions

Another challenge to Dalton's idea that one kind of atom can't change into another was the discovery of nuclear fusion and fission. Fusion is light elements getting together to make a heavier one, and fission is a heavy element splitting apart into lighter ones. For example, in the sun, hot gases combine:

$$_1^1 H + _1^2 H \rightarrow _2^3 He$$

And in the earth, radioactive solids decay:

$$_{92}^{238} U \rightarrow _{90}^{234} Th + _2^4 He$$

Why do they do this? It turns out there is a happy medium of nuclear stability (binding energy), at 55 to 60 mass units. Compared to this, H is small, so it fuses to get bigger. U is big, so it fissions to get smaller. Here's how nuclear stability changes with nuclear size:

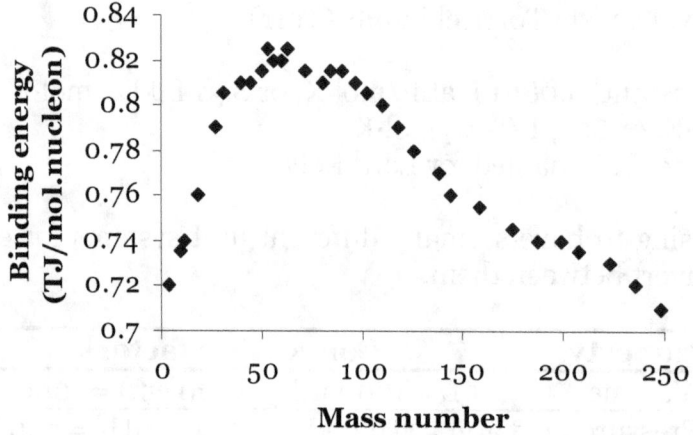

The nuclear binding energy can be calculated given a curious fact: the mass of an atom is less than the sum of its p^+ and n^0. In other words, some of the proton/neutron mass got converted to energy when they came together. Remember Einstein's equation:

E = (Δm)c²
E is the binding energy, in J or kg·m²/s²
Δm is the lost mass, in kg
c is the speed of light

The masses of the atomic particles are:

Particle	Mass in kg	Mass in amu or g/mol
Proton	1.673×10^{-27}	1.0073
Neutron	1.675×10^{-27}	1.0087
Electron	9.109×10^{-31}	0.00055

What is the binding energy, per nucleon, of helium-4, which weighs 4.0026 amu?

The isotope has 2 p^+ and 2 n^o, for a total of 4 nucleons, plus 2 e^-
Mass of 2 p^+ = 2(1.0073 amu) = 2.0146 amu
Mass of 2 n^o = 2(1.0087 amu) = 2.0174 amu
Mass of 2 e^- = 2(0.00055 amu) = 0.0011 amu
Total = 4.0331 amu

The actual mass of helium-4 is 4.0026 amu
Lost mass = 4.0331 − 4.0026 = 0.0305 amu
= 0.0305 g/mol = 3.05×10^{-5} kg/mol

$E = (\Delta m)c^2 = (3.05 \times 10^{-5}$ kg/mol$)(2.998 \times 10^8$ m/s$)^2$
= 2.741×10^{12} kg·m^2/s^2·mol = 2.741×10^{12} J/mol

Per nucleon: 2.741×10^{12} J/mol/4 nucleons = 6.9×10^{11} J/mol·nucleon
That's a lot of energy.

It also seems that isotopes with certain "magic" numbers of p^+ (2, 8, 20, 28, 50, 82, or 114) or n^o (2, 8, 20, 28, 50, 82, 126, or 184) are especially stable. So they too – like the e^-, whose magic (noble gas) numbers include 2, 8, 18, 36, 54, 86, and 118 – must have some special organizational structure.

And, there may an ideal balance of neutrons and protons (n/p ratio) that also guides the formation and decomposition of isotopes.

All of the heaviest elements in the periodic table are radioactive. Many are so unstable they can't be found in nature, and are only available synthetically. Once you make them, they decay spontaneously, releasing a lot of energy. This is the source of nuclear energy and weapons.

By contrast, as you can imagine from the sun example, fusion requires an input of energy and pressure that is hard or expensive for humans to create. Once it succeeds, there is a payoff in energy. The challenge with research into fusion is to find reactions with a reasonable cost and a good payoff. The search is motivated in part by the health dangers associated with fission.

Natural nuclear reactions follow certain patterns.

Alpha decay

A heavy isotope (Z > 83) ejects a He^{+2} ion, known as an alpha (α) particle, plus two electrons. For example:

$$^{238}_{92}\text{U} \rightarrow {}^{234}_{90}\text{Th} + {}^{4}_{2}\text{He}^{+2} + 2\,\text{e}^{-} \quad or \quad {}^{238}_{92}\text{U} \rightarrow {}^{234}_{90}\text{Th} + {}^{4}_{2}\alpha^{+2} + 2\,\text{e}^{-}$$

Beta negative decay (electron emission)

In a heavy or midweight isotope, a n^{o} turns into a p^{+}. The atomic number goes up by 1 unit, while the mass is unchanged, so the n/p ratio goes down. In the process, a beta particle ($_{-1}{}^{0}\beta$) is emitted, together with a subatomic particle called an antineutrino.

$$^{40}_{19}\text{K} \rightarrow {}^{40}_{20}\text{Ca} + {}^{0}_{-1}\beta + \text{antineutrino}$$

Beta positive decay (positron emission)

In a light isotope, a p^{+} becomes a n^{o}. The atomic number goes down 1, the mass is unchanged, the n/p ratio goes up, and a positron ($_{1}{}^{0}\beta$) is emitted together with a neutrino.

$$^{30}_{15}\text{P} \rightarrow {}^{30}_{14}\text{Si} + {}^{0}_{1}\beta + \text{neutrino}$$

Electron capture

A heavy or midweight isotope acquires a beta negative particle. In the process, the n/p ratio goes up and X-rays are emitted.

$$^{201}_{80}\text{Hg} + {}^{0}_{-1}\beta \rightarrow {}^{201}_{79}\text{Au} + \text{X-rays}$$

Gamma decay

A heavy isotope emits pure gamma (γ) energy. It appears to be unchanged, but of course something has gone on inside the nucleus.

$$^{222}_{86}\text{Rn} \rightarrow {}^{222}_{86}\text{Rn} + \gamma$$

Fission

A heavy isotope (Z > 90) splits up.

$$^{252}_{98}\text{Cf} \rightarrow {}^{140}_{54}\text{Xe} + {}^{108}_{44}\text{Ru} + 4{}^{1}_{0}\text{n}^{0}$$

The energy emitted by each of these, especially alpha and gamma, is so intense it can scramble your genetic material and provoke tumours, birth defects, and so on, given a high or frequent enough dose.

The product of a fission can itself be unstable and continue on. For example:

$$^{232}_{90}\text{Th} \rightarrow \alpha + {}^{228}_{88}\text{Ra} \rightarrow \beta + {}^{228}_{89}\text{Ac} \rightarrow \beta + {}^{228}_{90}\text{Th} \rightarrow \alpha + {}^{224}_{88}\text{Ra} \rightarrow \alpha + {}^{220}_{86}\text{Rn} \rightarrow$$

$$\alpha + {}^{216}_{84}\text{Po} \rightarrow \alpha + {}^{212}_{82}\text{Pb} \rightarrow \beta + {}^{212}_{83}\text{Bi} \rightarrow \alpha + {}^{212}_{84}\text{Po} \rightarrow \alpha + {}^{208}_{82}\text{Pb} \text{ (stable)}$$

Nuclear reactions can also be induced by humans. For example, an isotope can be bombarded with high energy particles, in a particle accelerator:

$$^{14}_{7}\text{N} + {}^{4}_{2}\text{He} \rightarrow {}^{17}_{8}\text{O} + {}^{1}_{1}\text{H}$$

$$^{14}_{7}\text{N} + {}^{1}_{0}\text{n}^0 \rightarrow {}^{14}_{6}\text{C} + {}^{1}_{1}\text{p}^+ + \text{e}^-$$

$$^{235}_{92}\text{U} + {}^{1}_{0}\text{n}^0 \rightarrow {}^{140}_{56}\text{Ba} + {}^{93}_{36}\text{Kr} + 3{}^{1}_{0}\text{n}^0$$

$$^{246}_{96}\text{Cm} + {}^{12}_{6}\text{C} \rightarrow {}^{256}_{102}\text{No} + 2{}^{1}_{0}\text{n}^0$$

Indeed, this is one way to make the elements that don't occur in nature. (The other main way is to capture the by-products of reactions run in nuclear power plants. And that's what holmium – the most naturally magnetic metal on Earth – is a specialist at!)

Why on earth would you want to make an element? One reason is to fill in a gap in the periodic table, and so learn more about the nature of matter. Another is for commercial, military, and other use, like americium in smoke detectors or plutonium in nuclear bombs.

The energy of a nuclear reaction can be calculated because it represents the mobilization of nuclear binding energy.

How much energy is released in the fusion of neon-20 into calcium-40, given that their respective binding energies are 1.28×10^{-12} and 1.37×10^{-12} J/nucleon?

$$^{20}_{10}\text{Ne} + {}^{20}_{10}\text{Ne} \rightarrow {}^{40}_{20}\text{Ca}$$

Total binding energy, initially:
(20 nucleons)(1.28×10^{-12} J/nucleon) + (20 nucleons)(1.28×10^{-12} J/nucleon)
= 5.12×10^{-11} J

Total binding energy, finally:
(40 nucleons)(1.37×10^{-12} J/nucleon)
= 5.48×10^{-11} J

Difference (final-initial):
$5.48\times10^{-11} - 5.12\times10^{-11} = 3.6\times10^{-12}$ J for one ^{40}Ne atom produced
(3.6×10^{-12} J/atom)(6.02×10^{23} atoms/mol) = 2.2×10^{12} J/mol of ^{40}Ne produced

Look how good the payoff is. But, we still haven't considered the initial cost. We'll do that in the chapter on kinetics.

Periodic trends in atom behaviour

The energy it takes to remove an e⁻ is called the ionization energy (IE), while the energy released when an atom gains an e⁻ is the electron affinity (EA). Since IE and EA change depending whether an element is in the solid, liquid, or gas state, by convention all elements are compared in the gas state.

An s^1 element like Na is happy to lose one e⁻ to become p^6, but doesn't want to lose a second e⁻ to leave that state, so its 1st IE is low while its 2nd IE is high. An s^2 like Mg will have low 1st and 2nd IE's and a high 3rd IE. A p^5 like F will have a high 1st IE because it would rather gain e⁻ (to get to p^6) than lose them. Here is the overall trend for the first few elements:

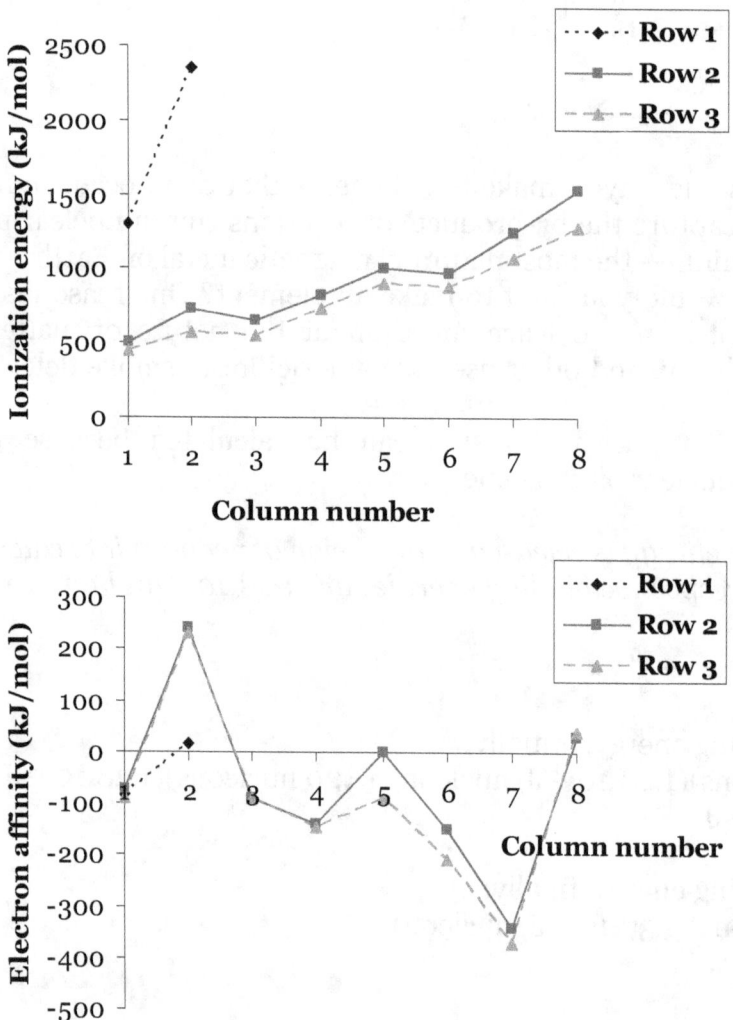

Look at the graphs and see if you can answer three questions.

Why, in general, does IE increase as you move from left to right in a row of the periodic table?

Why do IE and EA decrease when you jump down a row?

Why do they both go up in columns 2, 5, and 8?

If we go back to Bohr's equation of the energy of an e^- in its shell, we see that it depends on the number of p^+ present and how far the e^- is from the p^+. As you move across a row of the periodic table, you stay in the same shell, but the number of protons increases. Existing e^- are held more tightly, so they cost more IE to remove. New e^- get pulled in more strongly, so the EA is more favourable. When you jump down a row, you're a shell further away, so it gets easier to remove an e^- (the IE drops), and the attraction for a new e^- (EA) drops too.

Column 8 elements are very stable because they correspond to a full shell, with the valence p^6. s^2 (column 2), d^{10}, and f^{14} are quite stable because they're full subshells. p^3 (column 5), d^5, and f^7 are fairly stable because they are half-full, in other words they perfectly satisfy Hund's rule. This explains many of the oxidation states elements stop at along the way to p^6: C^{+2} (s^2), Zn^{+2} (d^{10}), and so on.

Another consequence of this is that, for some elements in neutral state, e^- will move further from the nucleus than they have to, to make a filled or half-filled configuration. For example:

You'd think Cr is ...$4s^2 3d^4$
but it's really ...$4s^1 3d^5$

Cu is not ...$4s^2 3d^9$
but ...$4s^1 d^{10}$

Pd is not ...$5s^2 4d^8$
but ...$4d^{10}$

If you're in school, your teacher can tell you which exceptions you need to memorize. Otherwise you can find them in a detailed periodic table.

These exceptions only work because the energetic cost of an e^- jumping away from the nucleus is offset by the benefit of a more stable final configuration. I said before that s and d happen to be close. Other subshells aren't, so transitions like the following one aren't worth it:

C is [He]$2s^2 2p^2$
not C [He]$2s^1 2p^3$

Since valence alone explains much of how an element behaves, we can usually neglect its inner (core) e- and replace that part of its configuration with the symbol of the corresponding noble gas. For example:

Na is $1s^2 2s^2 2p^6 3s^1$ *or* $[Ne]3s^1$
Se is $1s^2 2s^2 2p^6 3s^2 3p^6 4s^2 3d^{10} 4p^4$ *or* $[Ar]4s^2 3d^{10} 4p^4$

To a further approximation, only the s and p are needed to explain atoms' behaviour, so we distinguish these as true valence e- among the mass of outer shell electrons. For example:

Na has 10 core e- and 1 outer shell e-, of which 1e- is true valence
Se has 18 core e- and 16 outer shell e-, of which 6e- are true valence

Notice that the number of valence e- in a main-group atom can easily be found by looking at the periodic table. For example, column I elements (H, Li, etc.) have 1 valence e-, column II's (Be, Mg, etc.) have 2, and column VI's (O, S, etc.) have 6.

The size of an atom (atomic radius) is governed by the same p^+-e- attraction. It decreases as the attraction gets stronger, across a row of the periodic table, and increases as the attraction weakens, down a column.

The size of an ion (ionic radius) works much the same way. In my experience, exam questions tend to focus on comparing electronically equal (isoelectronic) ions of different elements.

Rank the following from smallest to largest size: S^{-2}, Cl^-, Ar, K^+.

S has $16p^+$, Cl has 17, Ar 18, and K 19
All have 18e-
So K^+ has the strongest p^+-e- pull, and is the smallest ion,
whereas S^{-2} has the weakest pull and is the largest ion
$K^+ < Ar < Cl^- < S^{-2}$

We say that elements like Na, who prefer to lose e- are relatively electropositive (δ^+), metallic, and electrically conducting, and combine with oxygen to make basic (alkaline) compounds, while those like F, who prefer to gain e-, are electronegative (δ^-), nonmetallic, and electrically insulating, and make acidic oxides.

In summary, here are the periodic trends of atomic behaviour:

Nonmetallic
Makes acidic oxides
Small atomic size
Strong proton-valence pull
Electronegative
High ionization energy
High electron affinity

Metallic
Makes basic oxides
Large atomic size
Weak proton-valence pull
Electropositive
Low ionization energy
Low electron affinity

Molecules:
The Buildings

Types of bonds

Once you know how an atom feels about its own valence e-, you know how it behaves toward another's. There are three main ways it can make a bond:

Metallic
Weak-pull (δ^+, metal) atoms hang out in a loose clump. For example, Rb with Rb, or Rb with Mg.

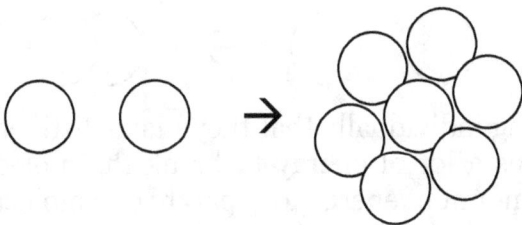

The bonds are weak individually and collectively. So these substances melt and boil at relatively low temperatures.

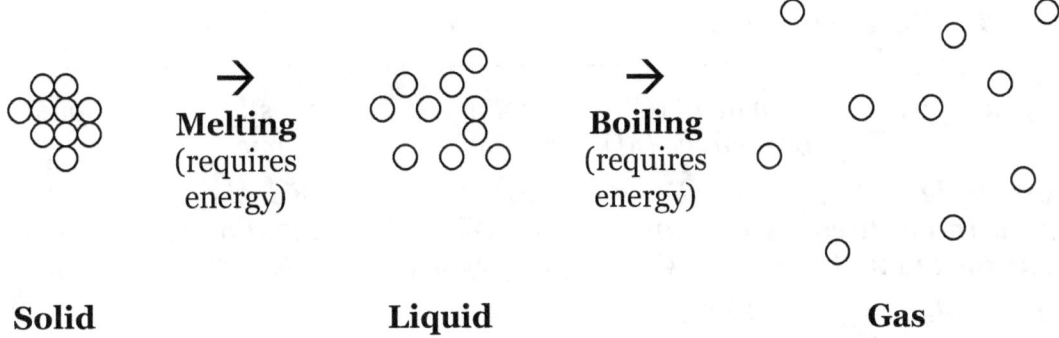

Solid   **Liquid**   **Gas**

And they tend to conduct electricity: valence e- are only loosely held, so free to roam.

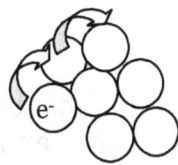

Ionic
A weak-pull atom meets a strong-pull (δ^-, nonmetallic) atom. The strong one takes an e- from the weak one; they become – and + ions, respectively; then they hang out in an organized crystal. For example, Rb with F.

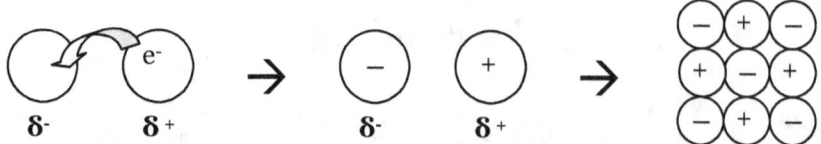

The bonds are weak individually, but strong collectively. So ionic compounds can dissolve in polar liquids, which have the ability to separate + and − ions, yet have high boiling points and melting points.

Covalent

Strong-pull atoms meet and have a tug-of-war. For example, F with F, or F with O.

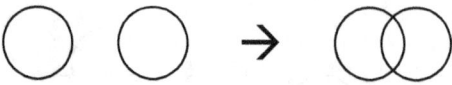

The bonds are so strong individually that they make distinct molecules, with variable collective forces. It takes a lot of energy to break the molecules back down to their atoms, and a variable amount of energy to separate one molecule from another.

A bond between two elements can be any mixture of these three archetypes. It depends on the difference in their electropositivity/electronegativity. Linus Pauling's scale ranks elements from 0.7 (most δ^+) to 4.0 (most δ^-).

$\delta^+ - \delta^-$ **difference**	0.0			3.3
bond type	covalent (C) or metallic (M)	polar covalent	mostly ionic	ionic (I)
example	F & F	F & C	F & Al	F & Fr
Pauling values	4.0, 4.0	4.0, 2.5	4.0, 1.6	4.0, 0.7
estimated nature	100% C	75% C, 25% I	35% C, 65% I	100% I
example	Fr & Fr			
Pauling values	0.7, 0.7			
estimated nature	100% M			

Metallic compounds in detail

Intro chemistry courses tend to say little about metallic compounds. In higher level courses, you can learn more about pure metals and their mixtures (alloys).

Ionic compounds in detail

Names and formulas

Ions exist as distinct entities only when dissolved in a liquid (solvent). Otherwise, + and − ions combine into solid crystals, where the total +'s and −'s are exactly balanced. By convention, we write the + ion first and the − ion second. For example:

1 Na^+ and 1 F^- combine to make NaF
2 Na^+ and 1 O^{-2} make Na_2O

Single-atom (monatomic) ions are named as follows:

○ + ions get the same name as the parent element;
○ − ions get the element's name plus the suffix "ide";
○ + ions of transition metals get the element's name, plus a roman numeral, since these elements don't have predictable charges.

For example:

Na^+ is "sodium"
F^- is "fluoride"
Cr^{+2} is "chromium(II)"
Cr^{+3} is "chromium(III)"

There are also some polyatomic ions:

CO_3^{-2} carbonate
HCO_3^- hydrogen carbonate ("bicarbonate")
$C_2O_4^{-2}$ oxalate
$CH_3CO_2^-$ acetate
ClO^- hypochlorite
ClO_2^- chlorite
ClO_3^- chlorate
ClO_4^- perchlorate
Hg_2^{+2} mercury(I)
MnO_4^- permanganate
NH_4^+ ammonium
NO_2^- nitrite

NO_3^- nitrate
O_2^{-2} peroxide
O_2^- superoxide
OH^- hydroxide
PO_3^{-3} phosphite
PO_4^{-3} phosphate
HPO_4^{-2} hydrogen phosphate ("biphosphate")
$H_2PO_4^-$ dihydrogen phosphate
SO_3^{-2} sulfite
SO_4^{-2} sulfate
HSO_4^- hydrogen sulfate ("bisulfate")

If you are in school, your teacher can tell you exactly which ones you need to know. And you can figure out others by association. For example, given that As is in the same column as P, AsO_3^{-3} is arsenite and AsO_4^{-3} is arsenate.

Here are some more examples of ionic compounds:

Mg^{+2} with ClO^- makes $Mg(ClO)_2$, magnesium hypochlorite
Ti^{+3} with S^{-2} makes Ti_2S_3, titanium(III) sulfide
NH_4^+ with O_2^- makes NH_4O_2, ammonium superoxide

NH_4^+ with O_2^{-2} makes $(NH_4)_2O_2$, ammonium peroxide
Li^+ with HCO_3^- makes $LiHCO_3$, lithium hydrogen carbonate

There is a special naming convention to use when your + ion is H.

- If the − ion ends in "ide", we call the whole compound "hydro (name of element) ic acid";
- If the − ion ends in "ate", it becomes "(name of element) ic acid";
- If it ends in "ite", we change it to "ous acid".

For example:

H^+ with S^{-2} (sulfide) makes H_2S, hydrosulfuric acid
H^+ with SO_4^{-2} (sulfate) makes H_2SO_4, sulfuric acid
H^+ with SO_3^{-2} (sulfite) makes H_2SO_3, sulfurous acid

According to Svante Arrhenius' definition, any molecule with H^+ behaves as an acid, while one with OH^- is basic. We'll see later on that there are other theories about this.

Predict and name a compound formed by:
a) Ca and Cl
b) N and C
c) Al and OH⁻
d) Mn(V) and SO$_4$⁻²

a) Ca is Ca^{+2}, Cl is Cl^-, making $CaCl_2$, calcium chloride
b) N^{+5} or N^{-3}, C^{+4} or C^{-4}, N is more electronegative than C, so N will be −3 and C will be +4, making C_3N_4, carbon nitride. However, this is a covalent compound, not an ionic one, since both are non-metals.
c) Al^{+3}, OH^-, $Al(OH)_3$, aluminium hydroxide

Name the following compounds.
a) Ba(NO$_3$)$_2$
b) HNO$_2$
c) Mn$_2$(SO$_4$)$_5$

a) Ba^{+2}, NO_3^-, barium nitrate
b) H^+, NO_2^- (nitrite), nitrous acid
c) we know SO4 is SO_4^{-2}, so Mn must be Mn^{+5}, making manganese(V)sulfate

Structures

Each ionic compound organizes itself in a characteristic structure (lattice) of repeating unit cells, depending on the size and charge of the ions involved.

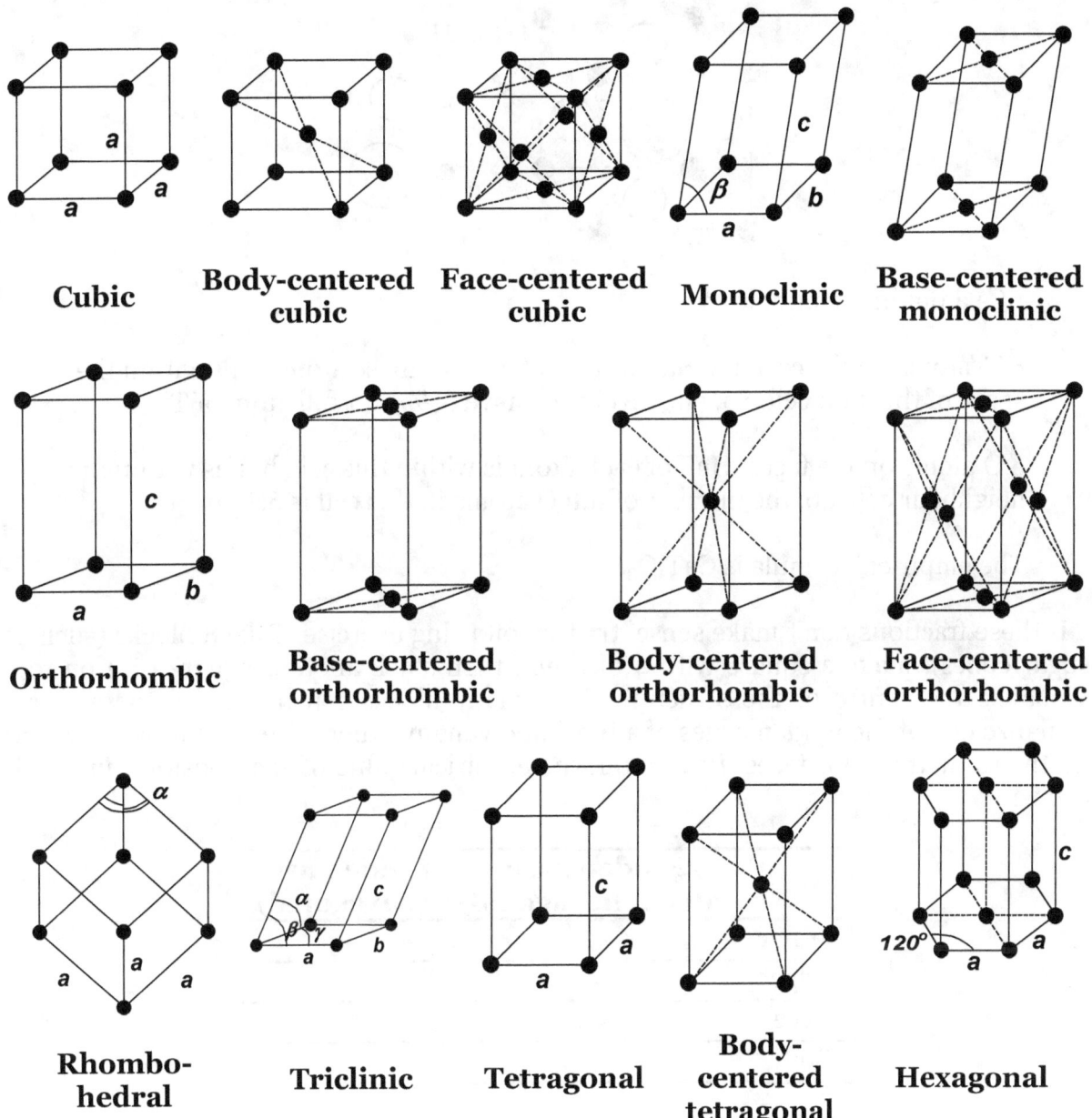

| Cubic | Body-centered cubic | Face-centered cubic | Monoclinic | Base-centered monoclinic |

| Orthorhombic | Base-centered orthorhombic | Body-centered orthorhombic | Face-centered orthorhombic |

| Rhombo-hedral | Triclinic | Tetragonal | Body-centered tetragonal | Hexagonal |

For each structure, we can determine the proportion (empirical formula) of the atoms involved.

What is the empirical formula of Perovskite, which has a cubic lattice with titanium on the corners, oxygen on the faces, and calcium in the centre?

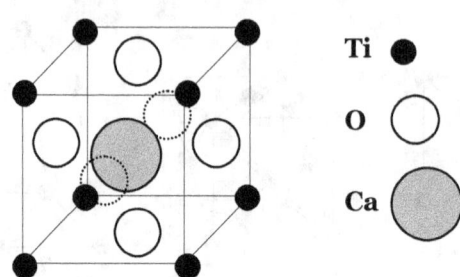

1 Ca atom in the middle

8 Ti atoms on the corners. But only ⅛ of each of these atoms falls within the space of this unit cell. So, effectively, we have 8(⅛) = 1 full atom of Ti

6 O atoms on the faces. Half of each atom is within this cell, half is within the neighbour cell. So, the number of full O atoms in this cell is 6(½) = 3.

The empirical formula is $CaTiO_3$.

   If these fractions don't make sense, try the following exercise. Take 8 blocks (such as sugar cubes), make a square with 4 of them, then stack another square of 4 on top. Visualize the centre of the structure: how many blocks touch corners there? Now visualize one of the hidden edges of a block: how many other edges is it touching? And do the same thing for faces. In general, the empirical value of each position in a unit cell is:

	4-sided base (all but hexagonal)	6-sided base (hexagonal)
Corner	⅛	⅙
Edge	¼	¼
Face	½	½
Body	1	1

Behaviour

We can also figure out how strongly the crystal is held together.

$U = N_A k q^+ q^- / r$
U is the bond energy, in J/mol
N_A is 6.022×10^{23} molecules/mol
k is a constant, 8.99×10^9 J·m/C²·molecule
q^+ is the charge of the + ion, multiplied by 1.61×10^{-19} C
q^- is the charge of the − ion, multiplied by 1.61×10^{-19} C
r is the distance between the centres of the + and − ions, in m

r is not always equal to the sum of the ionic radii, because they are not always sitting side-by-side in the crystal. Sometimes, you need trigonometry (geometric math) to calculate it.

The radius of an ion is the distance from its centre to its edge. This equals half its total width (diameter). The value depends somewhat on who the ion is paired with. You will either be given an average value or one specific to the crystal at hand.

What is the bond energy of $MgCl_2$ in a rhombohedral lattice, given that the cation's radius is 65 pm and the anion's is 181 pm.

The cation is Mg^{+2}
$q^+ = (2)(1.61 \times 10^{-19}\ C) = 3.22 \times 10^{-19}\ C$

The anion is Cl^-
$q- = 1.61 \times 10^{-19}\ C$

In a unit cell, the ions pack as tightly as they can. Generally, when you have only two kinds of ion in the unit cell, this means they touch. (Try drawing it.) So, here, the distance between the ions is the sum of their radii:
$65 + 181 = 246\ pm = 2.46 \times 10^{-10}\ m$

$U = N_A k q^+ q^-/r = (6.022 \times 10^{23}\ molecules/mol)(8.99 \times 10^9\ J \cdot m/C^2 \cdot molecule)(3.22 \times 10^{-19}\ C)(1.61 \times 10^{-19}\ C)/2.46 \times 10^{-10}\ m = 1.14 \times 10^6\ J/mol = 1140\ kJ/mol$

Now that's only for one bond. To measure the true lattice energy, you have to consider all bonds simultaneously.

Estimate the lattice energy.

Each cell has 4 Mg's in corners and 4 Cl's in opposite corners. In this way, an Mg and a Cl interact along each of the 12 edges of the cell. But, each edge is only ¼ in this cell, and ¾ in other cells. So, one cell has the equivalent of $(¼)(12) = 3$ interactions.

(3 bonds)(1140 kJ/mol·bond) = 3418 kJ/mol.

This doesn't match the actual lattice energy of $MgCl_2$, which is 2527 kJ/mol, because my reasoning was primitive: I considered Mg-Cl attractions but neglected Mg-Mg and Cl-Cl repulsions. Erwin Madelung to the rescue!

$U = M N_A k q^+ q^-/r$
U is now the lattice energy
M is the Madelung constant

Here are some sample values of M:

Compound	Unit cell type	Madelung constant
NaCl	Cubic	1.75
SiO_2	Hexagonal	2.22
TiO_2	Tetragonal	2.41
Al_2O_3	Rhombohedral	4.17

What is the Madelung constant for magnesium chloride?

2527 kJ/mol = (M)(1140 kJ/mol)
M = 2.22

Another way to find the lattice energy is to see how much energy is released when the independent (gas-phase) ions get together to make the compound. We'll do this in the chapter on thermodynamics.

What length is a Perovskite unit cell? The ionic radii are 114 pm for Ca^{+2}, 126 pm for O^{-2}, and 88 pm for Ti^{+4}.

Looking at those numbers and the drawing we made of the cell, earlier:
the cell's size is limited by the contact of Ca and O
(Ti and O don't touch, and neither do Ca and Ti)

Starting at one face, we cross half an O ion, a whole Ca ion, and half another O ion before arriving at the opposite face
So, we have travelled 1 O radius, 2 Ca radii, and 1 more O radius
126 + 2(114) + 126 = 480 pm

What is the volume?

The volume of a cube is length times length times length (length³)
$V = (480 \text{ pm})^3 = 1.11 \times 10^8 \text{ pm}^3$

What is the packing efficiency? In other words, how much of the cell's space is filled with ions?

The cell contains the equivalent of 1 Ca, 1 Ti, and 3 O ions
Each ion is a sphere
The volume of a sphere is $1.33\pi r^3$
$V_{Ca} = 1.33\pi(114 \text{ pm})^3 = 6.20 \times 10^6 \text{ pm}^3$
$V_{Ti} = 1.33\pi(88 \text{ pm})^3 = 2.85 \times 10^6 \text{ pm}^3$
$V_O = 1.33\pi(126 \text{ pm})^3 = 8.38 \times 10^6 \text{ pm}^3$
Total = $1(V_{Ca}) + 1(V_{Ti}) + 3(V_O) = 3.42 \times 10^7 \text{ pm}^3$

Packing efficiency = (total volume of ions/cell volume)100%
$= (3.42 \times 10^7 / 1.11 \times 10^8)100\% = 31\%$

Some sample lattice energies and melting points of ionic compounds are:

Compound	Ionic charges	Ionic radii (pm)	Lattice energy (kJ/mol)	Melting point (°C)
NaF	+1, -1	102, 133	911	902
NaCl	+1, -1	102, 181	788	801
$MgCl_2$	+2, -1	72, 181	2527	714
MgO	+2, -2	72, 140	3938	2700

You can predict whether or not an ionic compound will dissolve in water, using the following solubility rules:

- Compounds with column I (Na^+, K^+, etc.) ions always dissolve
- NH_4^+, NO_3^-, ClO_3^-, ClO_4^-, and $CH_3CO_2^-$ always dissolve
- Column VII (Cl^-, Br^-, etc.) ions dissolve, except when paired with Ag^+, Pb^{+2}, or Hg_2^{+2}
- SO_4^{-2} dissolves, unless paired with column II (Ca^{+2}, Sr^{+2}, etc.) or Pb^{+2}
- Anything else is only slightly soluble ("insoluble").

Many of the column VII and SO_4^{-2} compounds that don't dissolve in pure water can be made to dissolve if you acidify the water.

Decide if each of the following compounds is soluble or insoluble in water:
a) Ba(NO₂)₂
b) Li₃AsO₄
c) Cr₂S₃

a) Ba^{+2} and NO_2^- are both generally insoluble
b) Li^+ makes it soluble
c) Cr^{+3} and S^{-2} are both insoluble

Covalent compounds in detail

Names and formulas

These work much the same way as with ionic compounds. The δ^+ element precedes the δ^- element. Given that non-metals can have many different + ion states, they can combine with a − ion in more than one way. To say how many atoms of each are present, use prefixes:

1	mono	6	hexa
2	di	7	hepta
3	tri	8	octa
4	tetra	9	nova
5	penta	10	deca

If the δ^+ atom's prefix is mono, we usually drop it. Here are some examples:

CO carbon monoxide
CO_2 carbon dioxide
P_2S_5 diphosphorus pentasulfide

Structures

Covalent compounds are described using structures named for Gilbert Newton Lewis.

These are often taught using e^- dot diagrams, in which each atom's symbol is surrounded with its valence e^-, equally distributed on 4 sides. For example:

B, 3 valence e^-,   Ḃ.

C, 4 valence e^-,   .Ċ.

N, 5 valence e^-,   .N̈.

The 4 sides represent s, p_x, p_y, and p_z orbitals. A "lone" pair (LP) of e^- is stable as-is, while a single e^- (radical) is unstable and seeks to pair up with a single e^- from another atom. By convention, in the finished molecule, lone pairs are left as dots and bonds are shown as lines. For example, to make NH_3:

H̊

.N̈.   H̊   →   H:N̈:H   =   H–N̈–H
                              Ḧ |
H̊   H

Sometimes you need to make multiple bonds. For example, in O_2:

:Ö:Ö: →   :Ö::Ö: =   :Ö=Ö:

In my experience, this system works best for simple molecules. It can be amended to explain molecules with more than 4 things around them, like PH_5.

$$\ddot{.\!P.} \quad \rightarrow \quad \ddot{.\!P.} \quad \rightarrow \quad \begin{array}{c} \text{H} \quad \text{H} \\ \diagdown \; \diagup \\ \text{H–P–H} \\ | \\ \text{H} \end{array}$$

Essentially, the 5th side represents a d orbital, which we pretend is also valence. But now we have a problem: how do you know when to show an atom with 4 sides and when to show it with more?

That's why I've developed the following method, which works for all compounds. It may seem complicated at first. I go into such detail because Lewis structures (and acid-base theory) are the single most important general chemistry topic to master for success in organic chemistry, which comes next for many students. It should get easier with practice, and you'll have the time to do many examples, below.

I'll start with N_2O.

1. Count the e- available.
Total all atoms' valence e-. Remove 1e- for each + overall charge on the molecule. Add 1e- for each − overall charge on the molecule. It's helpful to distinguish different atoms of the same element, using subscripts.

N_1, 5 valence e-
N_2, 5 valence e-
O, 6 valence e-
no charge on molecule
Total, 16 e- available

2. Decide which atom is central.
The more δ^+ atom is central, because it is more willing to share its e-. H is never central, because it can only make 1 bond. The molecule doesn't have to be symmetrical.

N is central

3. Build a skeleton.
Make one bond between the central atom and each other atom. For oxoacids, which contain H, O, and one other element (for example, HNO_3 or H_2SO_4), attach H to O, not to the central atom.

$N_1–N_2–O$

4. Count the e- used so far.
1 bond is 2e-.

Made 2 bonds
Used 4 e-

5. Count the e⁻ remaining.

Had 16 e⁻
Used 4e⁻
12 e⁻ remaining

6. Total the e⁻ still needed by each atom to make it stable.

Most atoms want to have 8e⁻, like noble gases. This is the octet rule. There are many exceptions, though.

Column	Elements	Minimum e⁻	Maximum e⁻
I	H	2	2
II	Be	4	4
III	B, Al	6	8
IV	C, Si	8	8
V	N	8	8
	P, As	8	10
VI	O	8	8
	S, Se	8	12
VII	F	8	8
	Cl, Br, I	8	14
VIII	Kr, Xe	8	16

These are the only elements you need for Lewis structures because they are the only non-metals.

The pattern here is, most elements want to end up with a number of e⁻ equal to somewhere between 8 and twice their column number. Elements can only accept more than 8 e⁻ (expanded octet) starting in the third row of the periodic table, when they are big enough for the e⁻ containing groups around them to be far enough apart that they don't repel each other too much. Also notice that some noble gases can make bonds, after all.

N_1 wants 8e⁻, currently has 2e⁻, so needs 6e⁻ more
N_2 wants 8, has 4, needs 4 more
O wants 8, has 2, needs 6 more
Total, 6 + 4 + 6 = 16e⁻ more needed

7. Decide how to satisfy each atom using the e⁻ remaining.

○ If the #e⁻ needed = #e⁻ remaining, give each atom the e⁻ it wants, as LPs.
○ If #e⁻ needed < #e⁻ remaining, we have too many e⁻, so give each atom the LPs it wants, then put the extra e⁻ on an atom who can still accept more.
○ If #e⁻ needed > #e⁻ remaining, we don't have enough e⁻ to go around, so need to share. Make one more bond for every 2e⁻ too few, then put in LPs. Show all possibilities. You can put up to 3 bonds between any two atoms.

12e⁻ remaining
16e⁻ more needed
We have 4e⁻ too few
So, make 2 more bonds

$$N_1 \equiv N_2 - O \qquad N_1 = N_2 = O \qquad N_1 - N_2 \equiv O$$

Now put in LPs.

$$:N_1 \equiv N_2 - \ddot{O}: \qquad :\ddot{N_1} = N_2 = \ddot{O}: \qquad :\ddot{N_1} - N_2 \equiv O:$$

8. In each structure, see how satisfied or not each atom is, relative to its nature.

The tool for this is formal charge (FC). **FC = valence e⁻ - bonds – unbonded e⁻.**

N_1: 5 - 3 - 2 = 0 N_1: 5 – 2 – 4 = -1 N_1: 5 – 1 – 6 = -2
N_2: 5 – 4 – 0 = +1 N_2: 5 - 4 - 0 = +1 N_2: 5 – 4 – 0 = +1
O: 6 – 1 – 6 = -1   O: 6 – 2 – 4 = 0   O: 6 – 3 – 2 = +1

A FC of 0 represents stable balance. A +1 FC is okay on an δ⁺ atom but not an δ⁻ one (unless there are no alternative options). Likewise, a -1 FC is fine on an δ⁻ atom only. It's okay to have a +1 FC atom next to a -1 FC atom, but never a + next to a + or – next to –.

3rd structure is no good, because of + next to + and big (-2) FC
1st structure is best, because – FC is on more δ⁻ atom (O)
2nd structure is fine

9. See if the molecule can be improved by extra resonance.

Resonance means showing many possible structures for a given molecule. So, we have already shown some resonance. "Extra" resonance occurs in one situation only: you have a – FC atom with a LP next to a + FC atom who can accept more e⁻. In this case, use the LP to make an extra bond between the two atoms. You'll see that this brings both FC's to 0.

N_2O has a + atom next to a – atom with a LP, but the + atom (N_2) can't accept more e⁻, so there is no extra resonance

An example of extra resonance would be:

10. Report your answer, by giving all good resonance structures, from most to least stable.

$$N_2O = \left[\quad :N\!\equiv\!\overset{+}{N}\!-\!\overset{..}{\underset{..}{O}}\!:\overset{-}{} \quad \Longleftrightarrow \quad :\overset{-}{\overset{..}{N}}\!=\!\overset{+}{N}\!=\!\overset{..}{O}: \quad \right]$$

11. Give a time-averaged picture of these.
This is called a resonance hybrid.

$$N_2O = \left[\quad \overset{(-)}{N}\!\equiv\!\overset{+}{N}\!\cdots\!\overset{(-)}{O} \quad \right]$$

This picture shows that the N-N bond in this molecule is sometimes triple and sometimes double; the N-O bond is sometimes double and sometimes single; N_2 always has a + FC; and N_1 and O each have a − FC part of the time.

Resonance hybrids are useful for answering questions about bond length, bond strength, and charge. For example, if an N=N bond is 1.23 Å long (Å is the Anders Ångstrom unit, equal to 10^{-10} m) and N≡N is 1.09 Å, then the N-N bond in N_2O should be between 1.09 and 1.23 Å long, and indeed it is: 1.13 Å. Meanwhile, if the molecule meets up with a + charged ion, that ion will be attracted to O most of the time, because it has the − FC most of the time.

Find all good Lewis structures for:
a) NH_3
b) HCO_3^-
c) $XeOF_4$.

a) NH_3
N is central because H can't be
Total e^-: N $5e^-$, H $3\times1e^-$, no charge = $8e^-$

Skeleton:

$$H\!-\!\underset{|}{N}\!-\!H$$
$$H$$

We made 3 bonds, used $6e^-$, $2e^-$ left
N wants $2e^-$ more, H is content, $2e^-$ total missing
Put in lone pairs:

$$H\!-\!\overset{..}{N}\!-\!H$$
$$H$$

Formal charges:
N = 5 valence − 3 bonds − 2 unbonded e^- = 0
H = 1 − 1 − 0 = 0
No extra resonance is needed

No resonance hybrid

b) HCO_3^-
This is an oxoacid: C is central, H is attached to O
Total e^-: H 1, C 4, O 3x6, 1 extra (for $-$ charge) = 24

$$H-O_1-\underset{\underset{\textstyle O_3}{|}}{C}-O_2$$

Made 4 bonds, used $8e^-$, $16e^-$ left
H wants $0e^-$ more, C wants 2, O_1 4, O_2 6, O_3 6, total $18e^-$ missing
We have $2e^-$ too few available, so make 1 more bond, then put in lone pairs

$$H-O_1=\underset{\underset{\textstyle O_3}{|}}{C}-O_2 \qquad H-O_1-\underset{\underset{\textstyle O_3}{|}}{C}=O_2 \qquad H-O_1-\underset{\underset{\textstyle O_3}{\|}}{C}-O_2$$

$$H-\overset{..}{\overset{+}{O_1}}=\underset{\underset{\textstyle :\overset{..}{O_3}:}{|}}{C}-\overset{..}{\underset{..}{O}}_2: \qquad H-\overset{..}{\underset{..}{O}}_1-\underset{\underset{\textstyle :\overset{..}{O_3}:}{|}}{C}=\overset{..}{O}_2 \qquad H-\overset{..}{\underset{..}{O}}_1-\underset{\underset{\textstyle :\overset{..}{O_3}:}{\|}}{C}-\overset{..}{\underset{..}{O}}_2:^{-}$$

Formal charges:

H: 1-1-0=0	H: 1-1-0=0	H: 1-1-0=0
C: 4-4-0=0	C: 4-4-0=0	C: 4-4-0=0
O_1: 6-3-2=+1	O_1: 6-2-4=0	O_1: 6-2-4=0
O_2: 6-1-6=-1	O_2: 6-2-4=0	O_2: 6-1-6=-1
O_3: 6-1-6=-1	O_3: 6-1-6=-1	O_3: 6-2-4=0

The 2nd and 3rd structures are best, and equally good
No extra resonance is possible

$$HCO_3^- = \left[\quad H-\overset{..}{\underset{..}{O}}_1-\underset{\underset{\textstyle :\overset{..}{O_3}:^-}{|}}{C}=\overset{..}{O}_2 \quad \Longleftrightarrow \quad H-\overset{..}{\underset{..}{O}}_1-\underset{\underset{\textstyle :\overset{..}{O_3}:}{\|}}{C}-\overset{..}{\underset{..}{O}}_2:^- \quad \right]$$

Resonance hybrid:

$$HCO_3^- = \left[\quad H-O_1-\underset{\underset{\textstyle O_3\ (-)}{\vdots}}{C}\overset{(-)}{\cdots}O_2 \quad \right]$$

c) $XeOF_4$
Xe is central
Total e^-: Xe 8, O 6, F 4x7 = 42

```
      F  F
       \ /
    F–Xe–F
       |
       O
```

Made 5 bonds, used 10e⁻, 32e⁻ left
Xe wants 0e⁻ more, O 6, each F wants 6 more, total 30e⁻ missing
We have 2e⁻ too many available, so put in lone pairs, then dump extra e⁻ on Xe
who can take more

```
    :F: :F:
      \ /
   :F–Xe–F:
       | +
      :O:
       ‾
```

Formal charges:
Xe: 8-5-2=+1
O: 6-1-6=-1
F: 7-1-6=0

Extra resonance will happen:

```
    :F: :F:                    :F: :F:
      \ /                        \ /
   :F–Xe–F:         ⟺         :F–Xe–F:
       | +                        ‖
      :O:                        :O:
       ‾
```

All formal charges are now 0

  I mentioned before that all of the e⁻ containing objects (atoms and LPs, together known as coordination number or CN) around an atom in a Lewis structure repel each other. CN, then, determines the shape (orbital geometry) the groups will take, to minimize repulsion. This is known as valence shared electron pair repulsion (VSEPR) theory.

CN	Picture	Angles between bonds	Name of shape
1	A—X	-	Linear
2	X—A—X	180	Linear
3	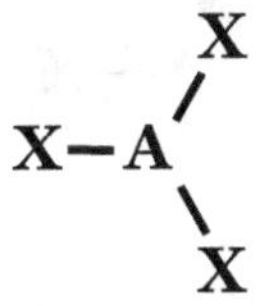 X—A with X above and X below	120	Trigonal planar
4	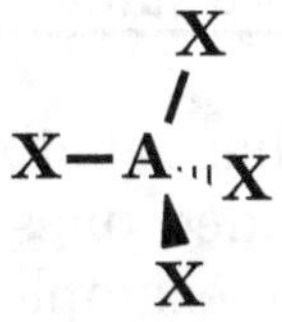 X—A⋯X with X above and X below	109	Tetrahedral
5	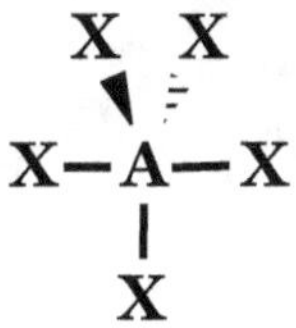 X X above, X—A—X, X below	90, 120	Trigonal bipyramidal
6	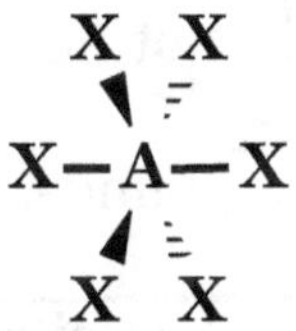 X X above, X—A—X, X X below	90	Octahedral

Here, a solid wedge represents something coming toward you, while a dotted wedge is going away from you. This is what happens when you try to deal with 3-dimensional objects on 2-dimensional paper!

In each shape, all positions around the central atom are equivalent, except in CN 5, where there are equatorial (eq) positions 120° apart from each other and axial (ax) positions 90° to either side of them.

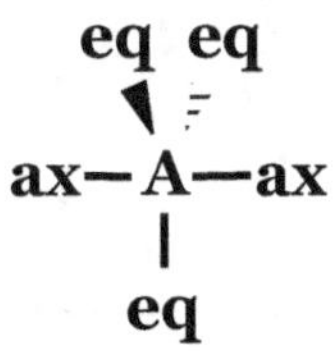

eq eq
ax—A—ax
eq

LPs repel more than atoms. For CN 5, this means we have to figure out which position they prefer, axial or equatorial. Looking at the worst repulsions only – in other words, the 90° ones, not the 120° ones:

Lone pair axial	Lone pair equatorial
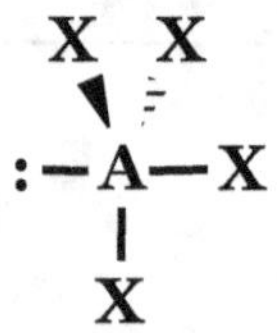	
3 LP-atom repulsions	2 LP-atom repulsions
Less stable	More stable

If you do the same exercise with 2 LP's, you'll see that equatorial is still most stable.

Triply bonded atoms repel more than double bonded ones, which repel more than singly bonded ones. This changes the angles a bit. For example:

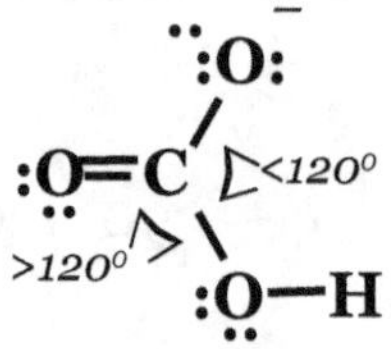

And LP's change it a lot. To get the true shape (molecular geometry) of a molecule with LPs on the central atom, show them right on it, not out in space, and show how they push the other atoms away.

The following chart covers all common covalent compounds.

Formula	Lewis structure	Atoms attached	Lone pairs	Total CN	Orbital Geometry	Molecular geometry
H_2	H—H	1	0	1	H—H Linear	H—H Linear
BeH_2	H—Be—H	2	0	2	H—Be—H Linear	H—Be—H Linear

BH_3	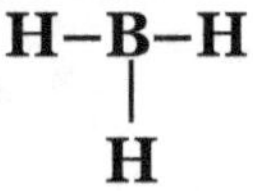 H—B—H
	     |
	     H
		3	0	3	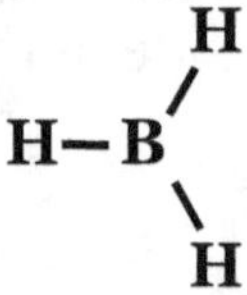 H—B (with H, H)
					 Trigonal planar
						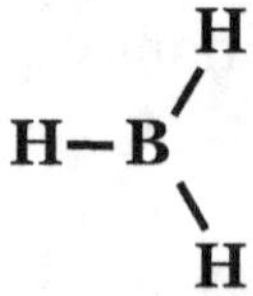 H—B (with H, H)
						 Trigonal planar

CH_2	H–C̈–H	2	1	3	:—C with H, H Trigonal planar	:C with H, H Bent
CH_4	H–C–H with H above and below	4	0	4	H–C with H, H Tetrahedral	H–C with H, H Tetrahedral
NH_3	H–N̈–H with H below	3	1	4	:—N with H, H Tetrahedral	:N with H, H Trigonal pyramidal
H_2O	H–Ö–H	2	2	4	:—O with H, H Tetrahedral	:Ö with H, H Bent
PF_5	F–P–F with F, F above and F below	5	0	5	F–P–F structure Trigonal bipyramidal	F–P–F structure Trigonal bipyramidal
SF_4	F–S–F with F, F above and lone pair below	4	1	5	F–S–F structure Trigonal bipyramidal	F–S–F structure Seesaw or Distorted tetrahedral

IF$_3$	F–I: (with F, F above)	3	2	5	F–I–F Trigonal bipyramidal	F–I–F T-shape
XeF$_2$	:Xe: (with F, F above)	2	3	5	F–Xe–F Trigonal bipyramidal	F–Xe–F Linear
SF$_6$	F–S–F (with F,F above and F,F below)	6	0	6	F–S–F Octahedral	F–S–F Octahedral
IF$_5$	F–I–F (with F above and F,F below)	5	1	6	:–I–F Octahedral	:I–F Square pyramidal
XeF$_4$	F–Xe–F (with F above and F below)	4	2	6	:–Xe–: Octahedral	:Xe: Square planar

In the AXE notation system, "A" stands for the central atom, "X" for the number of atoms attached, and "E" for the number of lone pairs. So, for example, NH$_3$ is AX$_3$E.

What are the orbital and molecular geometries of NH$_3$ and XeOF$_4$?

a) NH$_3$

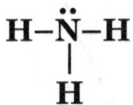

H–N̈–H
|
H

AX$_3$E
Orbital geometry: tetrahedral
Molecular geometry: trigonal pyramidal

b) XeOF$_4$

:F̈: :F̈:
\ /
:F̈–Xe–F̈:
‖
:O:

AX$_5$E
Orbital geometry: octahedral
Molecular geometry: square pyramidal

To understand bonds and LPs on a deeper level, we can look not just at the valence e⁻, but at the valence orbitals as well.

According to molecular orbital (MO) theory, to make a bond, the orbital on one atom has to match the symmetry of the orbital on the other atom. There are three main types of symmetry:

Symmetry	Symbol	Orbital relationship
Sigma	σ	Head to head
Pi	π	Side to side
Delta	δ	Face to face

By convention, in the 3-dimensional (x, y, z) system, head-to-head interactions are along the z axis.

Bonding (B) interactions are additive, stabilizing, and drawn by aligning the two contributing orbitals. Every time you make one, you also make a subtractive, destabilizing (antibonding, AB) interaction, shown by inverting the shading on one of the contributors. If you try to bring together two orbitals whose symmetries don't match, you get nonbonding (NB), on other words the orbitals remain unchanged.

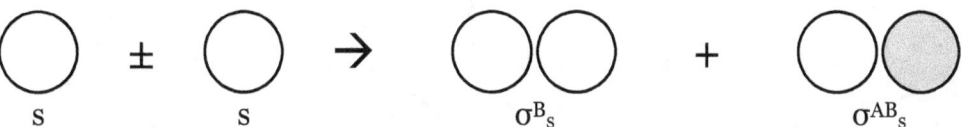

$$p_x \quad \pm \quad p_x \quad \rightarrow \quad \pi^B_{px} \quad + \quad \pi^{AB}_{px}$$

$$p_y \quad \pm \quad p_y \quad \rightarrow \quad \pi^B_{py} \quad + \quad \pi^{AB}_{py}$$

$$p_z \quad \pm \quad p_z \quad \rightarrow \quad \sigma^B_{pz} \quad + \quad \sigma^{AB}_{pz}$$

$$s \quad \pm \quad p_x \quad \rightarrow \quad \sigma^{NB}_s \quad + \quad \pi^{NB}_{px}$$

$$s \quad \pm \quad p_y \quad \rightarrow \quad \sigma^{NB}_s \quad + \quad \pi^{NB}_{py}$$

$$s \quad \pm \quad p_z \quad \rightarrow \quad \sigma^B_{s,pz} \quad + \quad \sigma^{AB}_{s,pz}$$

$$p_x \quad \pm \quad p_y \quad \rightarrow \quad \pi^{NB}_{px} \quad + \quad \pi^{NB}_{py}$$

$$p_x \quad \pm \quad p_z \quad \rightarrow \quad \pi^{NB}_{px} \quad + \quad \sigma^{NB}_{pz}$$

$$p_y \quad \pm \quad p_z \quad \rightarrow \quad \pi^{NB}_{py} \quad + \quad \sigma^{NB}_{pz}$$

In general chemistry, the first bond between two atoms is always σ. (This means that CN also = σ + LP.) The 2nd and 3rd bonds are π. Delta interactions are covered in more advanced courses. An example, just so you can visualize it, is:

$$d_{xz} \quad \pm \quad d_{xz} \quad \rightarrow \quad \delta^B_{dxz} \quad + \quad \delta^{AB}_{dxz}$$

Now, let's consider a full molecule, like CH_4. The structure is:

There are 4 bonds (4 σ, 0 π) and 0 LPs. The valence e⁻ configurations of C and H are:

C [He]$2s^2 2p^2$

H $1s^1$

So, we could imagine the bonds are formed as follows:

1st bond: C 2s with 1st H's 1s (σ)
2nd bond: C $2p_x$ with 2nd H's 1s (nonbonding)
3rd bond: C $2p_y$ with 3rd H's 1s (nonbonding)
4th bond: C $2p_z$ with 4th H's 1s (σ)

This says, CH_4 should have 2 σ bonds (↑) and 1 LP (↑↓). It also says the two σ bonds would be different lengths and strengths, since use pair different kinds of orbitals. Whereas, in reality, the CH_4 bonds are all the same. How do we fix this?

Hybridization says, for each atom, take all the orbitals it needs to make its σ bonds and LP, put them in a blender, and dish out however may pieces you put in. Leave out the orbitals needed for π bonds.

So, to hybridize C in CH_4, take the 2s, $2p_x$, $2p_y$, and $2p_z$ orbitals, put them in a blender, and dish out 4 portions. Each portion is ¼ s in nature and ¾ p. Since we used 1 s orbital and 3 p orbitals, we call it an $s^1 p^3$ (or sp^3) hybrid.

Hybrid orbitals look like a cross between the orbitals they come from. For this course, you can approximate all of them as follows (or even as just the bigger of the two lobes): .

C valence before hybridization

C valence after hybridization   ↑ ↑ ↑ ↑
                                 sp³ sp³ sp³ sp³
                                  σ σ σ σ

This now shows 4 bonds (all σ), and 0 LP, which is what we want. Notice that e⁻ can change spin in the process. Putting it all together, here is an orbital drawing of the molecule.

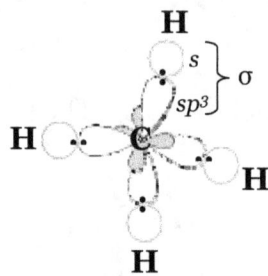

For many people, this is one of the most laborious parts of the course. Here are some more examples to make it more familiar.

H_2CO

:O=C⟨ H / H ⟩

C has 4 bonds (3 σ, 1 π), 0 LP
C's e⁻ configuration: [He]$2s^22p^2$ ↑↓   ↑ ↑ __
                                          s   p p p

CN = σ bonds + LP = 2 + 1 = 3
Therefore, hybridize 3 atomic orbitals (s, p, p), leaving over one p
C hybridized: ↑ ↑ ↑   ↑
                                sp² sp² sp²   p
                                 σ σ σ   π

Since O has many orbitals, we have to hybridize it too.

O has 2 bonds (1 σ, 1 π), 2 LP
O's e⁻ configuration: [He]$2s^22p^4$ ↑↓   ↑↓ ↑ ↑
                                          s   p p p

CN = σ bonds + LP = 1 + 2 = 3
Also sp² with p left over.
O hybridized: ↑↓ ↑↓ ↑   ↑
                                sp² sp² sp²   p
                                LP LP σ   π

Here is the whole molecule. Notice the lone pairs and the side-to-side (π) interaction of unhybridized p's.

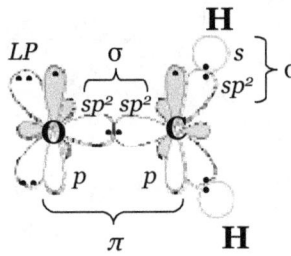

HCN

H−C≡N:

C has 4 bonds (2 σ, 2 π), 0 LP
CN = σ + LP = 2
sp hybrid
C unhybridized: [He]$2s^2 2p^2$

$\underset{\text{s}}{\uparrow\downarrow}$ $\quad$ $\underset{\text{p}}{\uparrow}$ $\underset{\text{p}}{\uparrow}$ $\underset{\text{p}}{\_\_}$

C hybridized:

$\underset{\underset{\sigma}{\text{sp}}}{\uparrow}$ $\underset{\underset{\sigma}{\text{sp}}}{\uparrow}$ $\quad$ $\underset{\underset{\pi}{\text{p}}}{\uparrow}$ $\underset{\underset{\pi}{\text{p}}}{\uparrow}$

N has 3 bonds (1 σ, 2 π), 1 LP
CN = 1 + 1 = 2
sp hybrid
N unhybridized: [He]$2s^2 2p^3$

$\underset{\text{s}}{\uparrow\downarrow}$ $\quad$ $\underset{\text{p}}{\uparrow}$ $\underset{\text{p}}{\uparrow}$ $\underset{\text{p}}{\uparrow}$

N hybridized:

$\underset{\underset{\text{LP}}{\text{sp}}}{\uparrow\downarrow}$ $\underset{\underset{\sigma}{\text{sp}}}{\uparrow}$ $\quad$ $\underset{\underset{\pi}{\text{p}}}{\uparrow}$ $\underset{\underset{\pi}{\text{p}}}{\uparrow}$

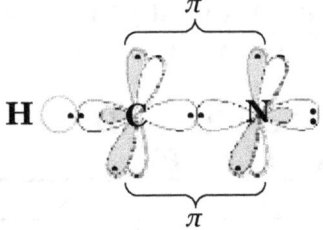

XeOF$_4$

:Ë: :Ë:
 \ /
:Ë–Xe–Ë:
   ‖
  :O:

Xe has 6 bonds (5 σ, 1 π), 1 LP

CN = σ + LP = 6

sp^{3}d^2 hybrid

Xe unhybridized: [Kr]5s^{2}4d^{10}5p^6 ↑↓   ↑↓ ↑↓ ↑↓ ↑↓ ↑↓   ↑↓ ↑↓ ↑↓

                                          s   d d d d d   p p p

This shows 0 bonds, 9 LP!

But we made a mistake: the hybrid uses 5s, 5p, and 5d (not 4d)

                                          ↑↓   ↑↓ ↑↓ ↑↓   __ __ __ __ __

                                          s   p p p   d d d d d

This shows 0 bonds, 4 LP

Xe hybridized: ↑↓ ↑ ↑ ↑ ↑ ↑   ↑ __ __

                                   all sp^{3}d^2 d d d

                                 LP σ   σ   σ   σ   σ   π   -   -

O is the same as in H$_2$CO, above (sp^2)

F has 1 bond (σ), 3 LP

CN = σ + LP = 4

sp^3 hybrid

F unhybridized: [He]2s^{2}2p^5 ↑↓   ↑↓ ↑↓ ↑

                                   s   p p p

F hybridized: ↑↓ ↑↓ ↑↓ ↑

                         sp^3 sp^3 sp^3 sp^3

The orbital picture for a molecule like this would be very crowded, and is more complicated than you will likely be asked to show.

You only have to go through all of this if you're specifically asked to show how the hybridizations are derived and draw the molecular orbitals. Otherwise, there's a shortcut:

CN	Total orbitals needed (σ, LP)	Hybrid	Left over (π)
1	1	s	-
2	2	sp	p, p
3	3	sp^2	p
4	4	sp^3	-
5	5	sp^{3}d	d, d, d, d
6	6	sp^{3}d^2	d, d, d

Now you can quickly and easily get the geometry and hybridization of any atom in any molecule, even a complicated-looking one.

Give the hybridization and geometry of each atom in the following compound:

$$\begin{array}{cc} \text{H} & \ddot{\text{O}}: \\ | & || \\ \text{H}-\text{C}_1-\text{C}_2-\text{C}_3\equiv\text{N}: \\ | \\ :\ddot{\text{F}}: \end{array}$$

Atom	σ bonds	Lone pairs	Total (CN)	Hybrid	Left over	Orbital geometry	Molecular geometry
C_1	4	0	4	sp³	-	Tetrahedral	Tetrahedral
H_1	1	0	1	s	-	Linear	Linear
H_2	1	0	1	s	-	Linear	Linear
F	1	3	4	sp³	-	Tetrahedral	Linear
C_2	3	0	3	sp²	p	Trigonal planar	Trigonal planar
O	1	2	3	sp²	p	Trigonal planar	Linear
C_3	2	0	2	sp	p, p	Linear	Linear
N	1	1	2	sp	p, p	Linear	Linear

Behaviour

According to the e⁻ dot model, every covalent bond comes from a perfect (nonpolar) sharing of e⁻ between two atoms. In reality, this is only true if the two atoms are the same, for example H and H. Any time the atoms are different – which is most of the time – the δ+ one gives more e⁻, the δ- atom takes more, and the bond is polar. We show this using a dipole moment arrow (vector).

$$\begin{array}{cc} \overset{\delta+}{} \qquad \overset{\delta-}{} & \overset{\delta+}{} \qquad \overset{\delta-}{} \\ \text{H}-\ddot{\text{F}}: & \text{H}-\ddot{\text{Br}}: \\ \longmapsto\!\!\rightarrow & \mapsto \end{array}$$

H-H, then, is said to have zero dipole moment (μ = 0 D, in units named for Peter Debye), and H-F has μ > 0. The bigger the electropositivity/electronegativity difference between the two atoms, the longer the vector. Lone pairs get the longest vectors of all.

To figure out the polarity of a molecule with many bonds and lone pairs, show a vector for each bond and lone pair, then add them. Vectors are added by putting them head-to-tail. If the vectors offset each other, the molecule is nonpolar, otherwise it's polar. In other words, if the molecule looks completely symmetrical, it's nonpolar, otherwise it's polar.

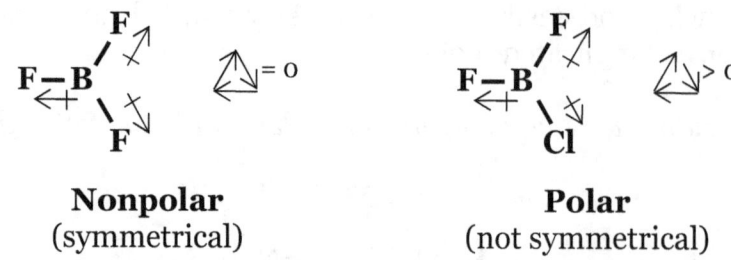

Nonpolar
(symmetrical)

Polar
(not symmetrical)

All molecules attract each other electrostatically, by temporary polarization of their e-clouds, for example in the presence of a strongly charged object. These "dispersion" forces are named for Fritz London and Johannes van der Waals. In general, the heavier a molecule is, the stronger its forces: more electrons and protons and greater size mean easier polarization.

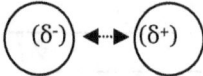

Polar molecules attract each other in these ways, plus by permanent dipole-dipole forces.

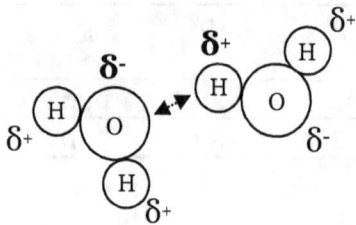

The more polar a molecule is, the stronger its dipole-dipole forces. If one molecule has an H attached to an N, O, F, or Cl (H-bond donor) and the other has a lone pair on an N, O, F, or Cl (H-bond acceptor), the force between the molecules is so strong it is distinguished as a "hydrogen-bond". Still, H-bonds are just dipole-dipole forces between molecules, not true bonds between atoms within a molecule. The last drawing is an example of H-bonding.

When you have a sample of a substance, you have millions of molecules of it. So even if the forces between any two molecules are much weaker than the bonds within one molecule, you can imagine how they start to add up. In general, the stronger the intermolecular forces, the harder it is to separate the molecules, so the higher the melting point (M_p) and boiling point (B_p). For instance:

Compound	Mass	Forces	M_p (°C)	B_p (°C)
CH_4	16	Dispersion	-182	-164
C_3H_8	44	Dispersion	-188	-42
CH_3Cl	50	Dispersion Dipole-dipole	-98	-24
CH_2Cl_2	85	Dispersion Dipole-Dipole	-97	40
$CHCl_3$	120	Dispersion Dipole-Dipole	-63	61
CHF_3	70	Dispersion Dipole-Dipole	-155	-82
CH_3OH	32	Dispersion Dipole-dipole H-bonding	-98	65

From this chart, we can see that:

O Mass affects B_p more than it does M_p (compare CH_4 and C_3H_8, or CH_3Cl and CH_2Cl_2);
O Polarity affects B_p and M_p less than mass does (compare $CHCl_3$ and CHF_3);
O H-bonding is the most powerful force (compare CH_3OH, CH_3F, and C_3H_8).

M_p and B_p are also affected by molecules' shapes. How would you make sense of the following data?

Molecular formula	Structure	M_p (°C)	B_p (°C)
C_5H_{12}	CH_3 / CH_2-CH_2 / H_3C-CH_2	-130	36
C_5H_{12}	CH_3 / $H_3C-C\cdots CH_3$ CH_3	-18	10

The rounder one packs tighter as a solid, so is harder to break up, and has the higher M_p. Imagine the amount of free space in a box of oranges versus a box of bananas. Meanwhile, the longer one has more surface area, which makes more attractive friction between molecules in the liquid state, so it is harder to boil.

In general, a nonpolar substance will dissolve better in another nonpolar substance than in a polar one, because of similar forces, and polars dissolve better in polars. In other words, "like dissolves like". A substance's ability to dissolve a polar compound is measured by its dielectric constant.

Heat also helps things dissolve. For example, sugar is easier to dissolve in tea or coffee than in cold water. And, as we have seen, acid or base can help some things dissolve too.

In-between compounds:
semiconductors, superconductors, and complexes

Semiconductors

In the chapter on atoms, we saw that their nature can be anywhere in a spectrum from totally metallic (conducting) to not at all metallic (insulating). The elements which are about half-and-half are called metalloids (semi-metals). Examples are Ga, Ge, and As. Many periodic tables show what looks like a staircase around which these elements cluster.

So, bonds between them are not clearly metallic, ionic, or covalent. This is where the next level of molecular orbital theory comes in handy.

If we took a simple covalent molecule like H_2, we could show the MOs in a diagram, like this:

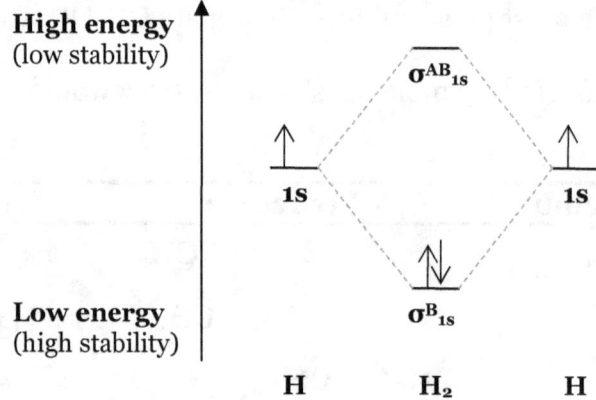

Moving to elements with more valence orbitals:

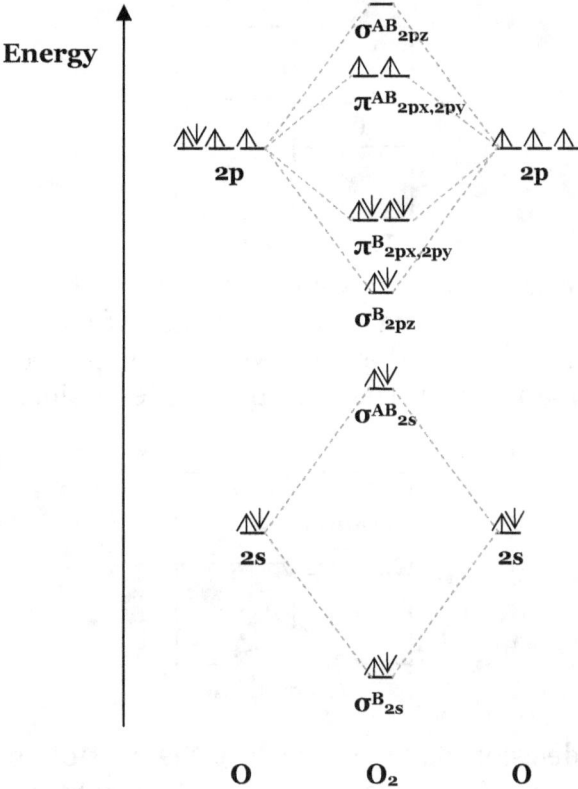

Notice a few things. A head-to-head (σ) orbital interaction is stronger than a side-to-side (π) one. There are 8 e⁻ in B orbitals and 4 in AB orbitals, for a net of 4 bonding e⁻, or 2 bonds, which agrees with the Lewis structure for O_2. This is how you calculate "bond order". And most of the orbitals contain e⁻.

If, instead of using O (which has 6 valence e⁻), we used C or Si (which have 4), the MOs would be exactly half-filled and half-empty. It could be represented like this:

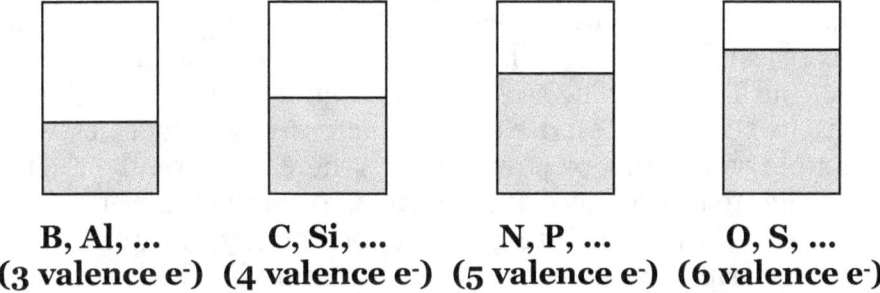

B, Al, ...   **C, Si, ...**   **N, P, ...**   **O, S, ...**
(3 valence e⁻) (4 valence e⁻) (5 valence e⁻) (6 valence e⁻)

Another way of seeing it is that C, Si, and column-mates are a halfway point between empty (0e⁻) and full valence (8e⁻), or between insulation and conductivity. They are called natural (intrinsic) semiconductors. Relative to them, B, Al, and mates are missing 1e⁻ ("holes", p-type), while N, P, and mates have an excess e⁻ ("plugs", n-type).

If you put an n-type material next to a p-type one, there is a powerful, one-way flow (current) of electrons.

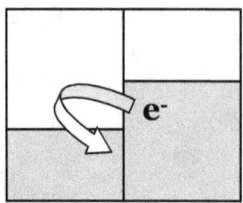

P-type　N-type

However, what is desired for electronics is a current that can be more finely controlled, and reversed by applying a small amount of energy. So, take a sample of intrinsic semiconductor and mix ("dope") it with a little p-type element. Take another sample and dope it with an n-type. Then put them side by side. Now you have the core of a transistor.

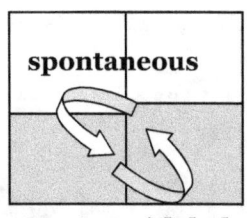

with help

This enables binary decision-making, which is the root of computer logic: follow a path until you get to a fork in the road, choose left or right, and continue along until you get to the next fork. "Left" and "right" are also known as "off" and "on", "0" and "1", and "p→n" and "n→p".

Superconductors

Some compounds are able to conduct electricity in a way that simple periodic trends and bonding models can't predict.

Actually, many metals conduct electricity perfectly, with no loss due to electrical friction (resistance), at temperatures below -250 °C. But that's not very useful in real life. So, there is research to develop "high-temperature" superconductors. So far, ones like $YBa_2Cu_3O_{\approx 7}$ and $Tl_2Ba_2Ca_2Cu_3O_{\approx 8}$ have been found to work as high as -150 °C. The eventual goal is to be able to send electricity across large distances with no loss of power, for example with trains or power lines. Whereas, currently, a lot is lost along the way, so more has to be produced at the source, to get the same effective output.

Notice that in the above structures the number of oxygen atoms is not an exact integer. That's what happens when you have a structure in which not all unit cells are the same.

Complexes

Because of the diffuse nature of d orbitals, transition elements also behave in ways that defy the simple categories of metallic, ionic, and covalent bonding.

We have seen that they tend to make + ions. So, they tend to be attracted to − ions. But, unlike regular ionic compounds, when these get together the charges don't necessarily have to equal zero. For example, Pt^{+2} and Cl^- make $PtCl_4^{-2}$. Transition metal cations can also be attracted to δ^- areas of neutral molecules, like Ni^{+2} with N in NH_3, to make $Ni(NH_3)_6^{+2}$.

These polyatomic ions of d block elements are called "complexes" or "coordination compounds", and their bonds are weak. They usually have the metal (M) in the middle, surrounded by 4, 5, or 6 "ligands" (L).

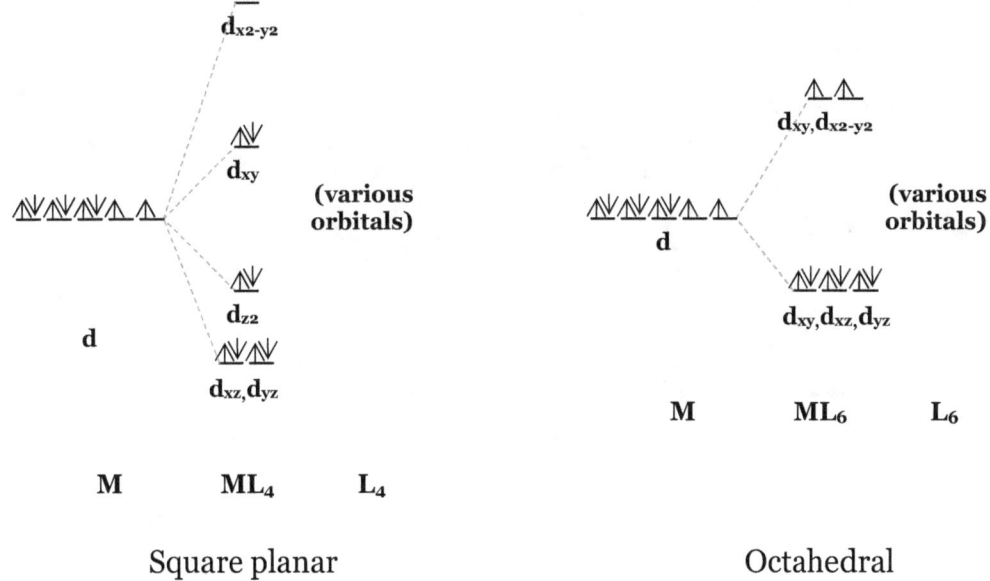

Common geometries are tetrahedral and square planar for CN 4, trigonal bipyramidal and square pyramidal for CN 5, and octahedral for CN 6. In each case, the ligands interact in a unique way with the metal's d orbitals. For example:

This can make for different magnetic behaviour. In the above examples, the square planar form is diamagnetic (all e⁻ are paired), while the octahedral form is paramagnetic (you have unpaired e⁻).

It can also change the amount of energy it takes for an e⁻ to jump from one orbital to the next highest one. This changes the colour of light emitted, and indeed one interesting feature of coordination compounds is their colour, arising from d orbital e⁻ jumps.

The spacing of the orbitals is also affected by the charge on the metal, the electronegativity of the atom in the ligand that associates to it, and whether or not that atom has π bonds.

The specific ways in which all of this happens are covered in higher inorganic chemistry. That's where you learn how to generate the MOs for all polyatomic compounds. It requires quite a bit of abstract math, known as "group theory".

Physical Processes
of Atoms and Molecules

Phases and phase changes

Intermolecular forces depend not just on a substance's nature, but on environmental conditions. Heat encourages molecules to spread out from one another, while cold brings them together. Pressure forces them close, whereas they expand to fill empty space (a vacuum).

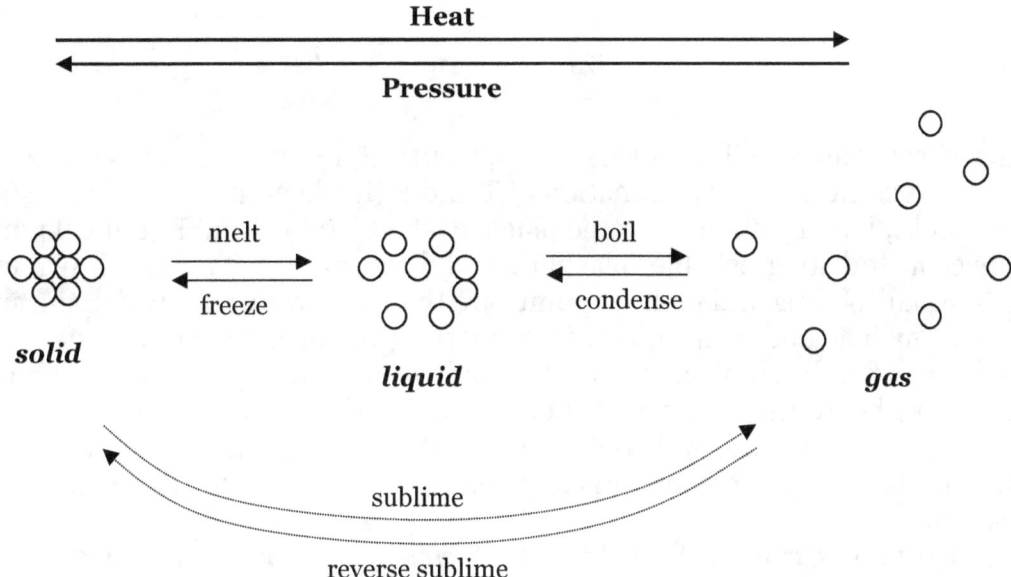

Solid (s), liquid (l), and gas (g), are the basic states (phases) of pure matter. Solids are fixed structures whose molecules can only vibrate. Liquids are flowing substances whose molecules can vibrate and rotate. Gas molecules can vibrate, rotate, and fly (translate).

Every substance changes phases over a certain range of temperatures (T) and pressures (P). Here's a typical phase diagram, for water:

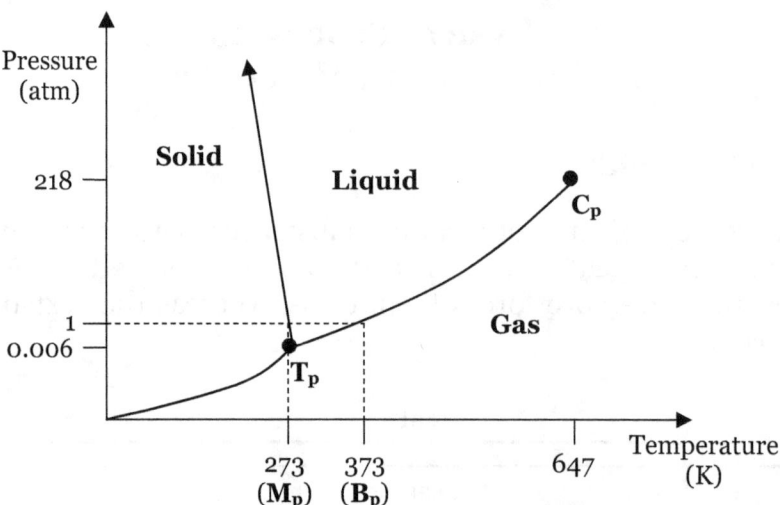

Notice a few things. As T increases, we go from s → l → g. As P increases, we go from g → l → s. At some special combination of T and P (triple point, T_p), for that substance, you can find s, l, and g all in the same place. Past a certain T and P (critical point, C_p), l and g become indistinguishable "plasma". And diagrams aren't always drawn to scale.

The "normal" boiling and melting points are the temperatures at which the substance boils and melts, under normal atmospheric pressure, which at sea level is 1 atmosphere. At a different pressure, the boiling and melting points will be different. Maybe you've heard that water boils below 100 °C when you're up on a tall mountain. Look at the phase diagram and you'll see why: pressure up there is below 1 atm. That's also why it's harder to breathe: air is less dense, so for the same volume inhaled in you get less substance.

Some substances can exist in different forms (allotropes), in a given phase. For example, carbon:

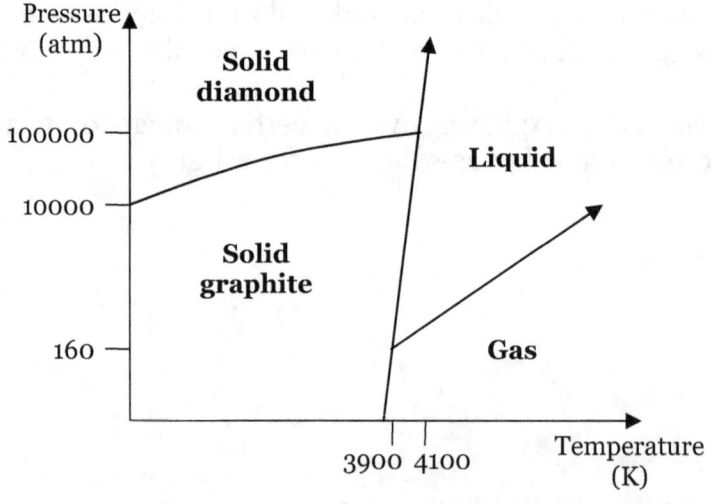

Graphite and diamond are both solids, but have different crystal structures because they are formed in different conditions of temperature and pressure.

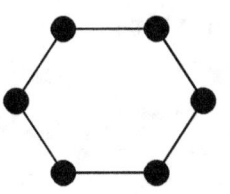 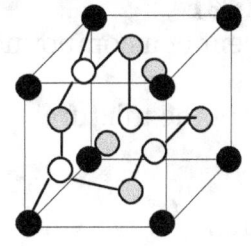

Graphite unit cell **Diamond unit cell**
(only some internal
bonds shown)

So, when we say that oxygen is a gas or silver is a solid, we're only talking about its state at standard temperature and pressure (STP), 1 atm and 25 °C. (Confusingly, in thermodynamics 0 °C is used.) Change the temperature and/or pressure, and you can make it any phase you like. But it might cost you a lot to build or run the equipment to do that. For example, look how much pressure it takes to make diamond, and imagine how you would muster it in real life!

Here are the standard states of the elements:

Gases	He, Ne, Ar, Kr, Xe, Rn, H_2, N_2, O_2, F_2, Cl_2
Liquids	Ga, Hg, Br_2
Solids	all other elements

As you can see, some elements naturally exist as isolated atoms, while others prefer a molecular form.

Mixtures

Even if each element has a standard state in nature, that doesn't mean it usually exists alone. For example, if you're lucky enough to find some vanadium – a useful ingredient in surgical stainless steel, among other things – it may be attached to oxygen, and the vanadium oxide may in turn be mixed in with other compounds, as in the ore "carnotite", $K_2(UO_2)_2(VO_4)_2 \cdot 3H_2O$.

- ○ A mineral is a mixture of solids, whose crystal structures are interwoven.
- ○ A solution is a solid, liquid, or gas ("solute") dissolved in liquid solvent. For example, salt water, vinegar water, or carbonated water.
- ○ A suspension is a solid that doesn't dissolve in liquid. For example, chalk in water.
- ○ An emulsion is a liquid that doesn't dissolve in liquid. For example, oil in water.

There are a few ways of measuring how much you have of each component.

m$_\%$ = (m/m$_T$)(100%)
m$_\%$ is the mass percent of a compound in a mixture
m is its mass
m$_T$ is the total mass of the mixture

m$_f$ = m/m$_T$
m$_f$ is the mass fraction

n$_\%$ = (n/n$_T$)(100%)
n$_\%$ is the mole percent of a compound in a mixture
n is its moles
n$_T$ is the total moles

X = n/n$_T$
X ("chi") is the mole fraction

M = n$_{solute}$/V$_{solution}$
M is molarity, in mol/L
n$_{solute}$ is the moles of solute
V$_{solution}$ is the volume of total solution, in L

Molarity is also called concentration, and shown in square brackets, "[ ]".

m = n$_{solute}$/m$_{solvent}$
m is molality, in mol/kg
n$_{solute}$ is the moles of solute
m$_{solvent}$ is the mass, in kg, of solvent

What is the molarity of 45% (by mass) alcohol (C_2H_6O) in water? The solution's density is 0.905 g/mL.

Assume 1 L of total solution
1 L = 1000 mL

(1000 mL)(0.905 g/mL) = 905 g solution
m of C_2H_6O = (45%)(905 g) = 407.2 g
n of C_2H_6O = 407.2 g/46.08 g/mol = 8.84 mol
M = n C_2H_6O/V solution = 8.84 mol/1 L = 8.8 mol/L

Solutions

Water-based (aqueous, "aq") solutions get studied in detail because they represent most of us and the earth's surface. Specifically, intro chemistry is interested in "colligative" properties, that is, behaviours that depend only on who the solvent is and how much solute you have, but not who the solute is.

For example, if you add some salt or sugar to water, it'll boil and freeze at a different temperature than the usual 100 °C and 0 °C. Which way do you think the temperature will change, higher or lower?

It turns out the boiling point goes up, because you introduce new intermolecular forces.

$$\Delta B_p = ik_Bm$$

ΔB_p is the change in B_p temperature, in °C
i is the number of particles per molecule of solute (Van't Hoff factor)
k_B is the solvent's boiling point elevation constant, in °C·kg/mol
m is the molality

i can be known theoretically or by calculation:

Determine the Van't Hoff factor for each of the following solutes:
a) table salt (NaCl)
b) sugar
c) potassium sulfate
d) hydrofluoric acid, in conditions under which it is 3.0% dissociated

a) NaCl is a soluble ionic compound, giving Na^+ and Cl^-, so i = 2
b) sugar is covalent, so i = 1
c) K_2SO_4 is a soluble ionic compound, giving $2K^+$ and $1SO_4^{2-}$, total i = 3
d) 3% of the HF exists as H^+ and F^- (i = 2), and 97% exists as HF (i = 1), so the average i is (3%)(2) + (97%)(1) = 1.06

To a 250 mL glass, you add a spoonful (35 g) of salt, fill to the top with water, and stir until dissolved. At what temperature will the contents boil? k_B of water is 0.52.

mol NaCl = 35 g/58.44 g/mol = 0.599 mol

g H_2O = (250 mL)(1 g/mL) = 250 g
kg H_2O = (250 g)(1 kg/1000 g) = 0.250 kg

molality of solution = 0.599 mol NaCl/0.25 kg H_2O = 2.40 mol/kg

i of NaCl = 2
ΔBp = (2)(0.52 °C·kg/mol)(2.40 mol/kg) = 2.5 °C
Pure water boils at 100 °C
This solution boils at 102.5 °C

Meanwhile, the freezing point (F_p) goes down when a solvent becomes impure: solute molecules get in the way of solvent molecules trying to pack together, so you have to lower the temperature to make them stay in place.

$\Delta F_p = ik_F m$
ΔF_p is the change in F_p temperature, in $^\circ C$
k_F is the solvent's freezing point depression constant, in $^\circ C \cdot kg/mol$

To a 250 mL glass, you add 30 g sugar ($C_6H_{12}O_6$), fill, and stir. At what temperature will the solution freeze, given k_F of water is 1.86?

mol sugar = 30 g/180.18 g/mol = 0.167 mol
molality = 0.167 mol sugar/0.25 kg H_2O = 0.666 mol/kg

i of sugar = 1
ΔF_p = (1)(1.86 $^\circ C \cdot kg/mol$)(0.666 mol/kg) = 1.2 $^\circ C$
Pure water freezes at 0 $^\circ C$
This solution freezes at -1.2 $^\circ C$

In general, solvents with stronger intermolecular forces – therefore higher B_p and M_p – have bigger K_B and K_F values.

Another way of saying that a solute makes it harder for a solvent to boil is, it makes the solvent evaporate less, or have a lower vapour pressure. Water is an example of a non-volatile liquid: standing at room temperature, it only slowly evaporates. Rubbing alcohol is an example of a more volatile liquid, with a higher vapour pressure. Want proof? Fill a glass with water and another one with rubbing alcohol, leave them out on the counter, and see how many days each takes to be gone.

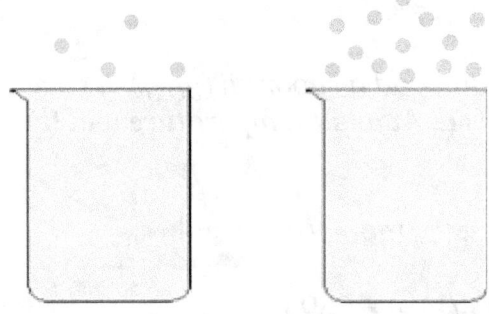

Non-volatile liquid   Volatile liquid

The vapour pressure of a solvent with a non-volatile solute is given by François-Marie Raoult's Law:

$P = P^*_{solvent}X_{solvent}$
$P^*_{solvent}$ is the vapour pressure of pure solvent
$X_{solvent}$ is the mole fraction of solvent in the whole solution

For the above salt solution, given that P^*_{H2O} is 23.8 mmHg at 25 $^\circ C$:

mol H_2O solvent = 250 g/18.02 g/mol = 13.87 mol
mol NaCl solute = 0.5988 mol
X_{H2O} = 13.87 mol/(13.87 mol + 0.5988 mol) = 0.9585

P = (23.8 mmHg)(0.9585) = 22.8 mmHg
The solute lowered the solvent's volatility.

In general, for a mixture of volatile or non-volatile liquids A, B, C, ...:

$P_{total} = P^*_A X_A + P^*_B X_B + P^*_C X_C + ...$

What is the vapour pressure of a solution of 50 g C_7H_8 (P^* = 3.79 kPa) and 80 g C_6H_6 (P^* = 12.68 kPa)?

nC_7H_8 = 50 g/92.15 g/mol = 0.543 mol
nC_6H_6 = 80 g/78.12 g/mol = 1.024 mol

X_{C7H8} = 0.543/(0.543 + 1.024) = 0.347
X_{C6H6} = 1.024/(0.543 + 1.024) = 0.653
Notice that the total is 1, or 100 %

$P_{total} = P^*_{C7H8} X_{C7H8} + P^*_{C6H6} X_{C6H6}$ = (0.347)(3.79 kPa) + (0.653)(12.68 kPa)
= 1.32 + 8.28 kPa = 9.6 kPa

Raoult's Law is an ideal model. In real life, the pressure of the mixture will be less than this, because the components have more intermolecular forces together than apart.

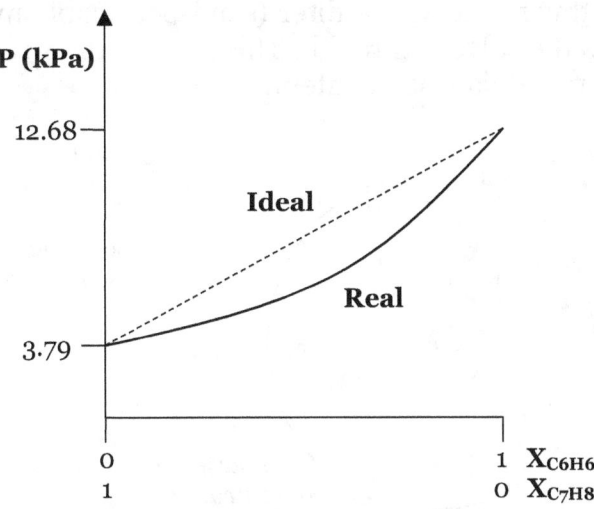

What is the mole fraction of each compound in the vapour phase?

The mole fractions we found before were in the liquid.

In the gas, given PV = nRT, P is proportional to n if V and T are constant. So:

$X_{C7H8} = n_{C7H8}/n_{total} = P_{C7H8}/P_{total}$ = 1.32 kPa/9.6 kPa = 0.14

$X_{C6H6} = n_{C6H6}/n_{total} = P_{C6H6}/P_{total}$ = 8.28 kPa/9.6 kPa = 0.86

Notice that C_6H_6 is even more highly represented in the gas than in the liquid, because it is more volatile.

Solutions also exert a kind of liquid (osmotic) pressure. To understand this, first consider simple diffusion.

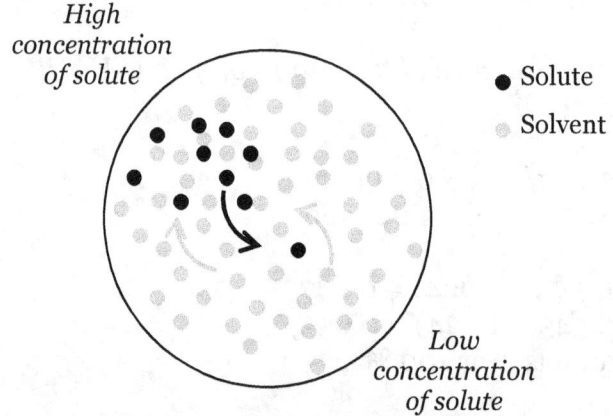

There is a natural tendency to even out the concentration everywhere. To do this, solute moves from the high concentration area to the low, and solvent moves in the opposite direction. Imagine putting a drop of blue food colour into a glass of water: you'll see the blue spread away, but what you may not see is the clear water rushing toward it. In the end, the whole solution will be evenly light blue.

Osmosis is the same thing, but with a filter (semi-permeable membrane). The filter stops the solute but lets the solvent through. This is what the wall of every cell in your body does, for example to regulate salt content.

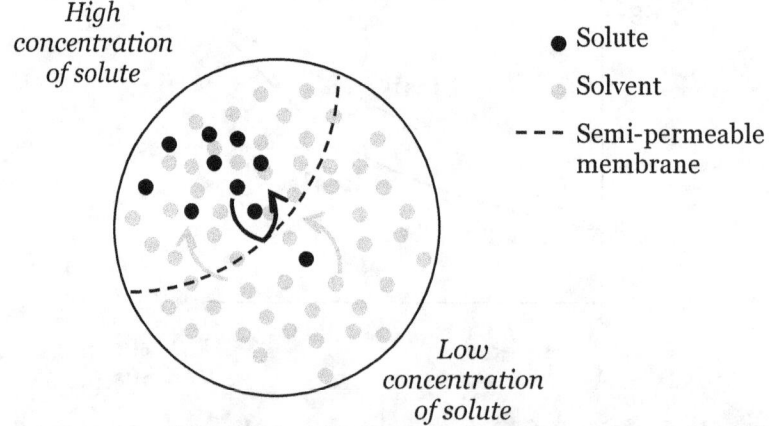

This means that volume will rise on the high-concentration side and drop on the low. For example, when you eat salty food, water rushes out of your cells, into your interstitial fluid, to try to dilute the salt, and your skin gets puffy. If left uncontrolled, the cells could become critically dry and/or what's outside could become critically swollen. To prevent this from happening, the body exerts pressure against the cell wall. This is osmotic pressure: the force you need to apply against the filter, to prevent solvent from leaking out.

π = iMRT
π is osmotic pressure, in atm, kPa, mmHg, or Torr
M is the solution's molarity
R is the gas constant
T is the temperature, in K

What is the molar mass of a protein that, at a concentration of 1.95 g per 100 mL of blood (ρ = 1.02 g/mL), exerts an osmotic pressure of 13.5 kPa at 37 °C?

Proteins are covalent, so i = 1
T = 37°C + 273 = 310 K
π = iMRT
13.5 kPa = (1)(M)(8.31 L·kPa/mol·K)(310 K)
0.005240 mol/L = M

per 100 mL (0.1 L) of solution:
(0.005240 mol/L)(0.1 L) = 0.0005240 mol protein

molar mass = 1.95 g/0.0005240 mol = 3720 g/mol
Indeed, proteins are generally large molecules!

Solids

Solids are fixed structures which can be described by crystal type, density, and other properties we have seen.

Liquids

Liquids are flowing structures. The forces between their molecules are still strong enough to give them some useful properties:

○ Surface tension: light solids dropped on them will float. For this reason, flies can walk on water.
○ Capillary action: liquids can climb the walls of thin tubes (capillaries), in defiance of gravity. This helps our blood circulate, for example.
○ Viscosity: they spread evenly. This is what motor oil does to the parts of an engine, to make sure there are no direct friction spots.

Gases

Gases get a lot of attention in general chemistry. Their behaviour is described by the kinetic molecular theory:

- ○ Low density: gas particles are tiny compared to the space between them
- ○ Volume: they spread out evenly to fill whatever container they are given
- ○ Pressure: they collide with the container's walls
- ○ Kinetic energy: the hotter the temperature is, the faster they move
- ○ Mass: the heavier they are, the slower they move

The average kinetic energy of a gas is:

KE = 3/2 nRT
KE is kinetic energy, in J/mol
n is moles
R is the gas constant, 8.31 J/mol·K
T is temperature, in K

The speed of gas molecules is measured as they travel through as a small opening. This is Thomas Graham's Law of effusion.

$$v = \sqrt{(3RT/M_M)}$$
v is the speed, in m/s
R is 8.31 J/mol·K = 8.31 kg·m²/s²·mol·K
M_M is the molar mass, in kg/mol

Graham's Law is often used to compare two gases, A and B.

given $v_A = \sqrt{(3RT_A/M_{M(A)})}$ and $v_B = \sqrt{(3RT_B/M_{M(B)})}$
$$v_A/v_B = \sqrt{(T_A \cdot M_{M(B)}/T_B \cdot M_{M(A)})}$$

Which gas, He or SO₂, effuses twice as fast as methane (CH₄) at 50°C?

The molar masses are:
He, 4 g/mol or 0.004 kg/mol
SO_2, 64 g/mol or 0.064 kg/mol
CH_4, 16 g/mol or 0.016 kg/mol

Since the gases are all at the same temperature:
$v_{He}/v_{CH4} = \sqrt{(M_{M(CH4)}/M_{M(He)})} = \sqrt{(0.016/0.004)} = 2$
$v_{SO2}/v_{CH4} = \sqrt{(M_{M(CH4)}/M_{M(SO2)})} = \sqrt{(0.016/0.064)} = 0.5$

He effuses twice as fast as CH_4, because it is 4 times lighter. SO_2 effuses half as fast, because it is 4 times heavier.

Robert Boyle observed that as P goes up, V goes down (gas is compressed), and as P goes down, V goes up (gas expands):

PV = constant
or, $P_{before}V_{before} = P_{after}V_{after}$
(often shown as $P_1V_1 = P_2V_2$)

Jacques Charles saw that as T goes up, V goes up (gas expands), and as T goes down, V goes down (gas shrinks):

V/T = constant
$V_{before}/T_{before} = V_{after}/T_{after}$

Joseph Gay-Lussac realized that as T goes up, P goes up, and as T goes down, P goes down.

P/T = constant
$P_{before}/T_{before} = P_{after}/T_{after}$

And no one took credit for noticing that as n goes up, P goes up (more molecules make more collisions), and as n goes down, P goes down.

P/n = constant
$P_{before}/n_{before} = P_{after}/n_{after}$

Each of these four equations assumes that the other variables (for example, with Boyle's Law, n and T) aren't changing. If you put them together, you get PV/nT = constant, and that constant is R. This is the ideal gas law, which we saw in the chapter on molecules.

What is the molar volume of a gas at 0 °C and 1 atm?

P = 1 atm
T = 0 °C + 273 = 273 K
n = 1 mol
PV = nRT
V = nRT/P = (1 mol)(0.0821 L·atm/mol·K)(273K)/(1 atm) = 22.4 L

This is pretty amazing: one mole of any gas behaves the same, pressure- and volume-wise, as a mole of any other gas. Therefore, for a mixture of gases A, B, C, ...:

$P_{total} = P_A + P_B + P_C + ...$

That's Dalton's Law of partial pressures. This comes in handy, for example, when running a reaction that produces a gas that is insoluble in water. It's hard to measure the volume of a gas directly. So, instead, let it come through a tube into a full container of water, and see how much water it displaces. However, when you measure its pressure, you have to separate out the part of the pressure that is due to the water.

> *You collect a gas over water at 20 °C, in a 5 L container. At this temperature, the vapour pressure of water is 2.3 kPa. The barometer reads 27.3 kPa. How many moles of the gas do you have?*

T in K = 20 °C + 273 = 293 K

$P_{total} = P_{gas} + P_{water}$
$P_{gas} = P_{total} - P_{water} = 27.3 - 2.3 = 25.0$ kPa

PV = nRT
n = PV/RT = (25.0 kPa)(5 L)/(8.31 L·kPa/mol·K)(293K) = 0.051 mol

Now actually, this is only an approximation, which is why we call it "ideal". In higher chemistry, you learn a more complex form of the equation, which adjusts P and V for the types of intermolecular forces specific to each gas.

Gases can be dissolved in liquids by pressuring them through. This, for example, is how you make carbonated water. The harder you push, the more dissolves. Open the can of soda, and the water falls flat, because the gas wants to be free.

The solubility of a gas in liquid is given by William **Henry's Law**:

s = kP
s is the solubility, in mol/L
k is the Henry's law constant for that gas, in L·atm/mol, L·kPa/mol, L·mmHg/mol, or L·Torr/mol.

The heavier the gas, the higher k tends to be. Its value does not depend on who the solvent is.

Chemical Processes
of Atoms and Molecules

Overview

In a physical process, such as a phase change, molecules stay intact. It's only the space and forces between them that change. In a chemical process, molecules get partly or totally broken down and/or reassembled. What you start with is called reagents or reactants, and what you end with is products. The state of each compound is usually shown.

There are a few common patterns:

Synthesis: A + B → AB

For example: $2H_{2(g)} + O_{2(g)} → 2H_2O_{(l)}$

Decomposition: AB → A + B

$PCl_{5(g)} → PCl_{3(g)} + Cl_{2(g)}$

Single displacement: A + BC → AC + B

$2Al_{(s)} + 6HCl_{(aq)} → 2AlCl_{3(aq)} + H_{2(g)}$

Double displacement: AB + CD → AD + CB

$HCl_{(aq)} + AgNO_{3(aq)} → HNO_{3(aq)} + AgCl_{(s)}$

Notice that the way the elements pair up depends on their relatively δ^+ or δ^- nature. For example, HCl is H^+ with Cl^-; when Al combines with Cl, Al is more δ^+, and its normal oxidation state is Al^{+3}, so the compound is $AlCl_3$.

These patterns are after-the-fact classifications. In other words, if someone tells you $2H_2 + O_2 → 2H_2O$, you can say it's a synthesis. But if they ask, $H_2 + O_2 → ?$, how can you predict what's going to happen? And where do those numbers (coefficients) in front of the molecules come from?

The answer is, get familiar with specific kinds of chemical reactions.

Combustion

In intro chemistry, combustion means a compound with C, H, N, O, and/or S reacts with O_2. Abundant O_2 and simple fuel (like propane, as opposed to tar) favour "complete" conversion of C to CO_2, H to H_2O, N to NO_2, and S to SO_3. Otherwise you may get some "incomplete" products products like CO, NO, and SO_2. Most exam questions assume complete combustion. So, if you're asked to complete the reaction:

$$C_3H_8 + O_2 \rightarrow ?$$

1. Show the products.

$$C_3H_8 + O_2 \rightarrow \quad CO_2 + H_2O$$

2. Add coefficients.
Each element must have as many atoms on the reagent side as on the product side. This is called balancing "by inspection".

Reagents	Products
C 3	C 1
H 8	H 2
O 2	O 3

To balance C:

$$C_3H_8 + O_2 \rightarrow \quad \mathbf{3}CO_2 + H_2O$$

C 3	C **3**
H 8	H 2
O 2	O 7

To balance H:

$$C_3H_8 + O_2 \rightarrow \quad 3CO_2 + \mathbf{4}H_2O$$

C 3	C 3
H 8	H **8**
O 2	O **10**

To balance O:

$$C_3H_8 + \mathbf{5}O_2 \rightarrow 3CO_2 + 4H_2O$$

C 3	C 3
H 8	H 8
O **10**	O 10

When you do it, you don't have to recopy each time. For example:

$$C_3H_8 + \mathbf{5}O_2 \rightarrow \mathbf{3}CO_2 + \mathbf{4}H_2O$$

C 3	C 1̶ 3
H 8	H 2̶ 8
O 2̶ 10	O 3̶ 7̶ 10

Sometimes, you may seem to get stuck with an even number of atoms of an element on one side and an odd number on the other.

$$2NH_3 + O_2 \rightarrow \quad 2NO_2 + 3H_2O$$

N 2   N 2
H 6   H 6
O 2   O 7

There are two ways out. One is to multiply the whole equation through by 2, to make all coefficients even. Then you can balance the last element.

$$4NH_3 + 2O_2 \rightarrow \quad 4NO_2 + 6H_2O$$

N **4**   N **4**
H **12**   H **12**
O **4**   O **14**

$$4NH_3 + 7O_2 \rightarrow \quad 4NO_2 + 6H_2O$$

N 4   N 4
H 12   H 12
O 14   O 14

Or, with diatomic elements only, you can use half-coefficients.

$$2NH_3 + \mathbf{3.5}O_2 \rightarrow \quad 2NO_2 + 3H_2O$$

N 2   N 2
H 6   H 6
O 7   O 7

Ionic displacement

An ionic compound reacts with another ionic compound and they switch partners. For example, given:

$$Na_3PO_4 + \quad BaCl_2 \rightarrow \quad ?$$

1. Show the reagents' ions.

Na^+ Ba^{+2}
$PO_4{}^{3-}$ Cl^-

2. Switch partners, to get the products.

Na^+ with Cl^- makes $NaCl$
$PO_4{}^{-3}$ with Ba^{+2} makes $Ba_3(PO_4)_2$

$$Na_3PO_4 + \quad BaCl_2 \rightarrow \quad NaCl + \quad Ba_3(PO_4)_2$$

3. Balance by inspection.
This gives the "molecular" equation.

$$2Na_3PO_4 + \quad 3BaCl_2 \rightarrow \quad 6NaCl + \quad Ba_3(PO_4)_2$$

You may be asked to stop here, or to continue on to the "total ionic" and "net ionic" equations. If so:

4. Label each compound as soluble (aq) or insoluble (s).
Use the solubility rules given in the chapter on molecules.

$$2Na_3PO_{4(aq)} + \quad 3BaCl_{2(aq)} \rightarrow \quad NaCl_{(aq)} + \quad Ba_3(PO_4)_{2(s)}$$

5. Break the soluble ones down to their ions.

$6Na^+$	$3Ba^{+2}$	$6Na^+$
$2PO_4^{-3}$	$6Cl^-$	$6Cl^-$

6. Recopy what's left.
This gives the total ionic equation.

$$6Na^+_{(aq)} + 2PO_4^{-3}{}_{(aq)} + 3Ba^{+2}{}_{(aq)} + 6Cl^-_{(aq)} \rightarrow 6Na^+_{(aq)} + 6Cl^-_{(aq)} + Ba_3(PO_4)_{2(s)}$$

7. Cancel spectator ions.
These are the ones that appear on both sides.

Na^+ and Cl^- are spectators

8. Recopy what's left.
This is the net ionic equation.

$$2PO_4^{-3}{}_{(aq)} + 3Ba^{+2}{}_{(aq)} \rightarrow Ba_3(PO_4)_{2(s)}$$

If, as in this example, you get a solid from two aqueous solutions, it's called a precipitate, and the reaction is a "precipitation".

Reduction-oxidation ("redox")

Here, one element gives electrons to another. They can be of the same element or different elements, and in the same compound or different compounds. For example:

$$MnO_4^- + Cl^- \rightarrow ?$$

It can be hard to predict the products of a redox reaction, so at least some of them will usually be given. For example:

$$MnO_4^- + Cl^- \rightarrow Mn^{+2} + Cl_2$$

1. Deduce the oxidation state of each element in each compound.
Start with the more predictable ones. Refer back to the atoms chapter if you need to.

$MnO_4^- + Cl^- \rightarrow Mn^{+2} + Cl_2$
in MnO_4^-, O should be -2, we have 4 O's for a total of -8, the total charge on the molecule is -1, so Mn has to be +7
in Cl^-, Cl is -1
in Mn^{+2}, Mn is +2
in Cl_2, Cl is 0 since there's no charge on the molecule

2. See who changes.

Mn: goes from +7 to +2, so it gains 5e⁻ (is reduced)
Cl: -1 → 0, loses 1e⁻ (oxidized)
O: -2 → ?

In redox reactions, the reagent containing the element being oxidized is called the reducing agent, and the one with the element being reduced is the oxidizing agent.

Mn is reduced, MnO_4^- is the oxidizing agent
Cl is oxidized, Cl^- is the reducing agent

3. Build "half-reactions".
For reduction, take the reagent and product containing the element being reduced. For oxidation, take the reagent and product containing the element being oxidized. From this point on, it helps to keep the reaction arrows lined up.

Reduction:
$MnO_4^- \rightarrow$ $\qquad\qquad$ Mn^{+2}

Oxidation:
$Cl^- \rightarrow$ $\qquad\qquad$ Cl_2

Don't worry about O's or H's that are missing on one side or the other. The method will take care of them soon. However, you have to make sure other elements are present on both sides. For example, given $H^+ + SO_4^{-2} + Zn \rightarrow ZnSO_4 + H_2$, for Zn's oxidation you can't just take $Zn \rightarrow ZnSO_4$; you have to show $SO_4^{2-} + Zn \rightarrow ZnSO_4$, to get the S's to balance too.

4. Balance the element being reduced or oxidized.

Red.:
$MnO_4^- \rightarrow$ Mn^{+2}

Ox.:
2$Cl^- \rightarrow$ Cl_2

5. Show e- being gained or lost.

Red.:
$MnO_4^- + \mathbf{5e^-} \rightarrow$ Mn^{+2}
We have 1 Mn atom, each Mn gains 5e⁻, so the total gain is 5e⁻

Ox.:
$2Cl^- \rightarrow$ $Cl_2 + \mathbf{2e^-}$
We have 2 Cl atoms, each Cl loses 1e⁻, total loss 2e⁻

For the next part, you have to know if the reaction is happening in an acidic or basic environment. This will be obvious from the compounds involved (for example, H_2SO_4 is acid), or you'll be told which to presume.

6. In acid, add H⁺ to balance overall molecular charges; in base, add OH⁻.

In acid:

Red.:
$MnO_4^- + 5e^- + \mathbf{8H^+} \rightarrow$ Mn^{+2}
We had -6 (from -1 and -5) on left, +2 on right, so we need +8 more on the left to make it +2 too

Ox.:
$2Cl^- \rightarrow$ $Cl_2 + 2e^-$
We had -2 on left and -2 on right, so didn't need to add anything

Or, in base:

Red.:
$MnO_4^- + 5e^- \rightarrow$ $Mn^{+2} + \mathbf{8OH^-}$
Both sides are now -6

Ox.:
$2Cl^- \rightarrow$ $Cl_2 + 2e^-$
The charges are already balanced

Many books teach a different method in base: add H^+, then add the same amount of OH^- to both sides and make H_2O on the side that had the H^+. I find this confusing, and have noticed that many of my students who learned it this way are less comfortable balancing redox reactions in base than in acid.

From here, the method is the same in acid or base. I'll continue with base.

7. Add H_2O to balance H and O.

Red.:
$$MnO_4^- + 5e^- + \mathbf{4H_2O} \rightarrow \quad Mn^{+2} + 8OH^-$$
We had 8 H's on the right, so need 8 H's on the left; the O's get taken care of in the process, as we now have 8 on each side

Ox.:
$$2Cl^- \rightarrow \quad\quad\quad Cl_2 + 2e^-$$
There are no H's or O's to balance

8. Make sure the number of e^- are the same in both half-reactions.
Multiply one or both of the equations through by some number, if necessary. We have to do this because one substance can only gain as much as the other loses!

2 x Red.:
$$\mathbf{2MnO_4^- + 10e^- + 8H_2O} \rightarrow \quad \mathbf{2Mn^{+2} + 16OH^-}$$

5 x Ox.:
$$\mathbf{10Cl^- \rightarrow} \quad\quad\quad \mathbf{5Cl_2 + 10e^-}$$

9. Add the two half-reactions.
This makes a redox reaction. In the process, the e^- cancel.

$$2MnO_4^- + 10e^- + 8H_2O \rightarrow \quad 2Mn^{+2} + 16OH^-$$
$$\underline{10Cl^- \rightarrow \quad\quad\quad\quad\quad 5Cl_2 + 10e^-}$$
$$2MnO_4^- + 8H_2O + 10Cl^- \rightarrow \quad 2Mn^{+2} + 16OH^- + 5Cl_2$$

10. Check that it's balanced.

Mn 2	Mn 2
O 16	O 16
H 16	H 16
Cl 10	Cl 10
charge -12 (from -2 and -10)	charge -12 (from +4 and -16)

It can happen that some molecules of a compound are reduced while others are oxidized. For example:

$Cl_2 \rightarrow Cl^- + ClO_2^-$
in Cl_2, Cl is 0
in Cl^-, Cl is -1
in ClO_2^-, Cl is +3

Some Cl atoms are reduced from 0 to -1
Some Cl atoms are oxidized from 0 to +3

Red.: $Cl_2 + 2e^- \rightarrow 2Cl^-$
Ox.: $Cl_2 + 4H_2O \rightarrow 2ClO_2^- + 8H^+ + 6e^-$

3 x Red.: $3Cl_2 + 6e^- \rightarrow 6Cl^-$
Ox.: $Cl_2 + 4H_2O \rightarrow 2ClO_2^- + 8H^+ + 6e^-$

Redox: $4Cl_2 + 4H_2O \rightarrow 6Cl^- + 2ClO_2^- + 8H^+$

Empirical and molecular formulas

One way this knowledge of reactions is used is to deduce the structure of a compound from the products of its reaction.

A compound is known to contain only carbon, hydrogen, and oxygen, but the proportion is unknown. When a 0.45 g sample of it is combusted, 0.64 g of CO_2 and 0.39 g of H_2O are recovered. What is the proportion of elements in the compound?

mol CO_2 = 0.64 g/44.01 g/mol = 0.0145 mol
This contains 0.0145 mol C
mass C = (0.0145 mol)(12.01 g/mol) = 0.175 g

mol H_2O = 0.39 g/18.02 g/mol = 0.0216 mol
This contains (2)(0.0216) mol H = 0.0433 mol H
mass H = (0.0433 mol)(1.01 g/mol) = 0.0437 g

The combustion was: compound + $O_2 \rightarrow CO_2 + H_2O$
All of the C in the compound went into CO_2
All of the H went into H_2O
The O went into both places, so is hard to track
We had 0.45 g compound
Of this, 0.175 g were C, 0.044 g were H, and the rest (0.45 − 0.175 − 0.044 = 0.231 g) were O
mol O = 0.231 g/16.00 g/mol = 0.0144 mol

At this point, you can find the percent composition of the compound, by mass, if asked.

% C = (0.175 g C/0.45 g total)(100%) = 39%
% H = (0.044 g H/0.45 g total)(100%) = 10%
% O = (0.231 g O/0.45 g total)(100%) = 51%
Notice that this adds up to 100%.

To find the ratio between a set of numbers, divide each one by whoever is smallest.

mol C/mol O = 0.0145 mol/0.0144 mol = 1.01 ≈ 1
mol H/mol O = 0.0433 mol/0.0144 mol = 3.01 ≈ 3
mol O/mol O = 0.0144 mol/0.0144 mol = 1
The compound has C, H, and O in a proportion of 1 to 3 to 1
In other words, the empirical formula is $(C_1H_3O_1)_n$

The n says, we don't yet know how many of these units the molecule has. It could be CH_3O just as well as $C_2H_6O_2$, or $C_3H_9O_3$, and so on.
Sometimes, when you do this last step, you won't get perfect integers. Let's say you got a ratio of 1 to 2.5 to 1. You can't have 2.5 atoms of H in a molecule! But, you can multiply all the subscripts by whatever number gets rid of the decimals: 2 x $C_1H_{2.5}O_1$ = $C_2H_5O_2$

If the compound's molar mass is known to be 62 g/mol, what is its molecular formula?

One empirical unit (CH_3O) of the compound weighs (1)(12.01) + (3)(1.01) + (1)(16.00) = 31.04 g/mol

The total mass is 62 g/mol, which is equal to 2 empirical units:
n = molar mass/empirical mass = 62/31.04 = 2.0 ≈ 2

The molecular formula is $(CH_3O)_2$, or $C_2H_6O_2$.

Another way these questions are asked is to skip the combustion data and give the percent composition directly. This is less work.

A compound contains, by mass, 39% carbon, 10% hydrogen, and the rest oxygen. What is its empirical formula?

The % of oxygen is 100 - 39 - 10 = 51 %

We need to assume some amount of compound. 100 g is convenient.
g C = (39%)(100 g) = 39 g
g H = (10%)(100 g) = 10 g
g O = (51%)(100 g) = 51 g

From here, all you have to do is convert each of these to moles and take the ratio, as above.

Stoichiometry

What we just did is an example of stoichiometry: relating the amount of reagents and products in a chemical reaction. No matter what kind of reaction you have and how the data are given, relate them by moles.

A → 2B

Says 1 molecule of A gives 2 molecules of B
or, 1 mol A gives 2 mol B

But not, 1 g A → 2 g B
or, 1 mL A → 2 mL B
and so on.

For example, to find how many grams of hydrogen you can get from the decomposition of 15 g NH_3:

1. Give a balanced chemical equation.

$2NH_3$ → N_2 + $3H_2$

2. Get moles for whichever compound you have the fullest data on.

mol NH_3 = 15 g/17.04 g/mol = 0.880 mol

3. Relate reagent to product.

2 mol NH_3 gives 3 mol H_2
So, mol H_2 = (3/2)(0.880 mol) = 1.320 mol

4. Report the answer in the units asked for.

g H_2 = (1.320 mol)(2.02 g/mol) = 2.7 g

If there are many reagents, you have to figure out which one to relate to the products. Imagine making something from separate ingredients, like a banana-strawberry smoothie. Let's say, for 1 mug of smoothie, you need 2 bananas and 9 strawberries.

2B + 9S → 1M

Now let's say you have 5 bananas and 32 strawberries to start with. How many mugfuls can you make?

   For 1 mug, use 2B and 9S
   For 2 mugs, use 4B and 18S
   For 3 mugs, use 6B and 27S
   For 4 mugs, use 8B and 36S

We have enough strawberries for 3 mugs, but only enough bananas for 2, so we can only make 2 smoothies (unless we change the recipe!). In other words, banana is the limiting reagent, and the amount of product depends only on it.
Sometimes you'll be told which reagent is limiting.

   38.3 g CH4 are burned in excess oxygen. How much water should be produced?

   $CH_4 + 2O_2 \rightarrow CO_2 + 2H_2O$

   mol CH_4 = 38.3 g/16.04 g/mol = 2.388 mol

   CH_4 is limiting
   mol H_2O = (2)(mol CH_4) = (2)(2.388 mol) = 4.78 mol

Other times, you have to calculate which one is limiting.

   38.3 g CH_4 are reacted with 100 L oxygen at 24°C and 1.05 atm. How much water should be produced?

   mol CH_4 = 38.3 g/16.04 g/mol = 2.39 mol
   mol O_2 = PV/RT = (1.05 atm)(100 L)/(0.0821 L·atm/mol·K)(24+273 K) = 4.31 mol

Don't jump to the conclusion that CH_4 is limiting because there's less of it! We have to check if we have less or more in proportion to the amount of O_2 needed. Do this by dividing each reagent's number of moles by its balancing coefficient:

   2.39 mol CH_4/1 = 2.39
   4.31 mol O_2/2 = 2.15

Whoever gives the smaller result is limiting.

   O_2 is limiting
   mol H_2O = (2/2)(mol O_2) = (2/2)(4.31 mol) = 4.3 mol

   How much reagent is left over?

CH_4 is left over
We had 2.39 mol
We used $(1/2)(mol\ O) = (1/2)(4.31\ mol) = 2.15\ mol$
Left over = 2.39 − 2.15 = 0.24 mol

If 5.3 g H_2O were recovered when this reaction was carried out, what was the percent yield?

Theoretically, we should have gotten:
(4.31 mol)(18.02 g/mol) = 77.7 g

Percent yield = (actual/theoretical)(100%)
= (5.3 g/77.7 g)(100%) = 6.8%

In general, a yield close to 100% is a sign that the experiment was well done, and a low yield suggests it wasn't. There can be many reasons: some of the product was spilled before being measured; the reaction wasn't given time to finish; and so on.

How many mL of 0.16M NaOH are needed to neutralize 180 mL of 0.25 M $Ca(NO_3)_2$?

$2NaOH_{(aq)} + Ca(NO_3)_{2(aq)} \rightarrow 2NaNO_{3(aq)} + Ca(OH)_{2(s)}$

volume $Ca(NO_3)_2$ solution = 180 mL/1000 mL/L = 0.180 L
mol $Ca(NO_3)_2$ = (0.18 L)(0.25 mol/L) = 0.0450 mol

mol NaOH needed = (2)(0.0450) = 0.0900 mol
volume NaOH = 0.0900 mol/0.16 mol/L = 0.562 L = 560 mL (to 2 sig figs)

What mass of precipitate should be obtained?

Both reagents are limiting, so either can be related to the product

mol $Ca(OH)_2$ = (1/1)(mol $Ca(NO_3)_2$) = (1/1)(0.0450 mol) = 0.0450 mol
g $Ca(OH)_2$ = (0.0450 mol)(74.1 g/mol) = 3.3 g

What is the concentration of each ion left in solution?

Both reagents were used up, so the only aqueous species now present is $NaNO_3$

Mol $NaNO_3$ = (2/1)(mol $Ca(NO_3)_2$) = (2/1)(0.0450 mol) = 0.0900 mol
This contains 0.0900 mol Na^+ and 0.0900 mol NO_3^-

The total volume is 0.180 + 0.562 = 0.742 L
The concentrations are:

Na^+ 0.0900 mol/0.742 L = 0.12 mol/L or 0.12 M
NO_3^- is also 0.12 M

You are given a mixture of sodium and calcium oxides, totaling 4 g, and asked to determine the percentage of each component in it. When you dissolve the oxides in water, they turn into their respective hydroxides, totaling 5.25 g.

The components of a mixture don't react with each other, but simply coexist. So it would be wrong to write:

NaO + CaO + H_2O →

Instead, show a separate reaction for each component:

$Na_2O + H_2O$ → 2NaOH
CaO + H_2O → $Ca(OH)_2$

It can make things clearer to create a variable representing the mass of each component:

mass of Na_2O = a
mass of CaO = b
Therefore, a + b = 4 g

mol Na_2O = a/61.98 g/mol = (1/61.98)a = 0.01613a
mol CaO = b/56.08 g/mol = (1/56.08)b = 0.01783b

mol NaOH produced = 2(mol Na_2O) = 2(0.01613a) = 0.03227a
mol $Ca(OH)_2$ produced = 1(mol CaO) = 1(0.01783b) = 0.01783b

mass NaOH = (0.03227a mol)(40.00 g/mol) = 1.291a g
mass $Ca(OH)_2$ = (0.01783b mol)(74.10 g/mol) = 1.321b g

The total mass of product is 5.25 g
Therefore, 1.291a + 1.321b = 5.25

Now we have a system of 2 equations in 2 variables:
a + b = 4
1.291a + 1.321b = 5.25

To solve, rearrange the first equation to isolate one variable:
a = 4 − b

Plug this in to the other equation:
1.291(4-b) + 1.321b = 5.25

$5.163 - 1.291b + 1.321b = 5.25$

Isolate the other variable:
$-1.291b + 1.321b = 5.25 - 5.163$
$0.03b = 0.087$
$b = 0.087/0.03 = 2.9$ g CaO

Therefore:
$a = 4 - b = 4 - 2.9 = 1.1$ g Na_2O

The percentage of CaO in the mixture is:
(2.9 g CaO/4 g total)(100%) = 72%
That leaves 28% Na_2O

Equilibrium:
Reversible Processes

Overview

Some reactions go and never come back. Like burning a piece of paper: imagine trying to collect all the ashes, gases, and other products, and put them back together! Other processes can be reversed. For example, running an electrical current (electrolysis) through water to break it up into its elements:

$$2H_2O \rightarrow 2H_2 + O_2$$

If you collect the H_2 and O_2 together and light a match, you'll get water back:

$$2H_2 + O_2 \rightarrow 2H_2O$$

The short way of showing this is:

$$2H_2O \Leftrightarrow 2H_2 + O_2$$

There is a forward process and a reverse process, and depending on what external conditions you supply, one of three things will happen:

○ The forward process dominates, depleting reagents and accumulating products
○ The reverse process dominates, reagents go up and products go down
○ The two processes are in exact balance (dynamic equilibrium), so the amount of reagents and products stays the same

Le Chatelier's principle

Henri Louis Le Chatelier found that every reversible process wants to get to equilibrium, and that if it gets there and you disturb it, it will try to undo whatever you did, to get back. For example, given:

$$A_{(g)} + 3B_{(s)} + 2C_{(aq)} \Leftrightarrow 4D_{(g)} + 2E_{(l)} + heat$$

If the system is at equilibrium and you add some A, the system wants to use up A, and it does this by favouring the forward reaction, until it gets back to equilibrium.
If you remove some A, it wants to make more, so the reverse reaction dominates.
If you add some D, it wants to use it up, so goes backward.
If you remove some D, it wants to make more, so goes forward.
If you add or remove a liquid or solid (like E or B), equilibrium is not disturbed.

And if you add or remove a substance, like a noble gas, that is not involved in the reaction, equilibrium is not disturbed.

Why don't solids and liquids matter? Equilibrium means, for a given reaction, there is a certain stable proportion of products to reagents. This constant (K) can be measured in terms of concentration (C) or pressure (P).

K_c = [products]/[reagents]
K_p = $P_{products}/P_{reagents}$

You can only measure the concentration of a (g) or (aq) species, and you can only measure the pressure of a (g).

Each compound's concentration or pressure is raised to the power of its balancing coefficient. For the above reaction, this means:

$K_c = [D]^4/[A][C]^2$
$K_p = P_D^4/P_A$

Given PV = nRT, or P = nRT/V, and given that C = n/V, or P = CRT:

$K_p = K_c(RT)^{\Delta n}$
Δn = total coefficients of gas products – total of gas reagents

In the above example, Δn = 4 – 1 = 3

Le Chatelier's Principle also works for changes to the overall reaction conditions. Continuing with the above example:

If you raise the temperature (T), that is like adding heat, so the system wants to use up heat, and to do that it goes backward. If you lower T, it wants to make heat, so goes forward.

If you raise the pressure (P), it wants to reduce P, so it favours whichever direction consumes more gas than it makes – in this case, the reverse. If you lower P, it wants to raise P, so makes more gas, going forward.

If you increase the volume (V), that decreases the P, so it goes forward. If you lower V, it goes backward.

If you add a catalyst, equilibrium is not disturbed, because a catalyst speeds up the forward and reverse reactions equally.

General set-up for calculations

1-way reactions are what we have already done with stoichiometry: they stop when you run out of one reagent.

	A +	2B →	C
Initial amount (I)	3 mol	5 mol	0 mol
Change (C)	-2.5 mol	-5 mol	+2.5 mol
Final amount (F)	0.5 mol	0 mol	2.5 mol

2-way reactions go only partway in one direction, leaving a balance of all compounds. For example, using K_c:

	$A_{(aq)}$ +	$2B_{(g)}$ ⇔	$C_{(aq)}$
Initial concentration (I)	3 mol/L	5 mol/L	0 mol/L
Change (C)	-x	-2x	+x
Equilibrium concentration (E)	3-x	5-2x	x

$$K_c = [C]/[A][B]^2 = (x)/(3-x)(5-2x)^2$$

Given a value for x, you can solve for K_c. For example:

In the above reaction, the equilibrium concentration of C is found to be 0.52 M. What is the value of K_c?

$[C] = x = 0.52$ M
$[A] = 3 - x = 3 - 0.52 = 2.48$ M
$[B] = 5 - 2x = 5 - (2)(0.52) = 3.96$ M
$K_c = (0.52$ M$)/(2.48$ M$)(3.96$ M$)^2 = 0.013$ $1/M^2$

What is K_p, in atm, at 40 °C?

T in K = 40 + 273 = 313
R = 0.0821 L·atm/mol·K
$\Delta n = 0 - 2 = -2$
$K_p = K_c(RT)^{\Delta n} = (0.0134)[(0.0821)(313)]^{-2} = 2.0 \times 10^{-5}$ $1/atm^2$

Likewise, given a value for K_c, you can solve for x. For example:

If $K_c = 3.7 \times 10^{-7}$, what is the equilibrium concentration of C?

$3.7 \times 10^{-7} = (x)/(3-x)(5-2x)^2$

This is a cubic equation, which is hard to solve. In such cases, you are usually allowed to make a simplifying assumption:

Since $K_c \ll 1$, [products] « [reagents], so x « 3 or 2x « 5, and:
$3.7\text{x}10^{-7} \approx (x)/(3)(5)^2$
$(3.7\text{x}10^{-7})(3)(5)^2 \approx x$
$2.8\text{x}10^{-5} \approx x$
$[C] = x \approx 2.8\text{x}10^{-5}$ M

Typically, the assumption is considered valid if the x you get is less than 5 or 10 % of the number it was paired with. So, for example:

Assume x < (5%)(3) = (0.05)(3) = 0.15 which is true!

As long as K is small, the assumption tends to work. If K is big, we have to do something funny: assume the reaction goes all the way, then comes back a bit.

If $K_c = 3.7\text{x}10^7$, what is the equilibrium concentration of C?

	$A_{(aq)}$ +	$2B_{(g)}$ ⇔	$C_{(aq)}$
Initial concentration (I)	3 mol/L	5 mol/L	0 mol/L
Change (C)	-2.5	-5	+2.5
Temporary concentration (T)	0.5	0	2.5
Change (C)	+x	+2x	-x
Equilibrium concentration (E)	0.5+x	2x	2.5-x

$K_c = [C]/[A][B]^2$
$3.7\text{x}10^7 = (2.5-x)/(0.5+x)(2x)^2$

Assume x « 2.5 and x « 0.5
$3.7\text{x}10^7 \approx (2.5)/(0.5)(2x)^2$
$(3.7\text{x}10^7)(2x)^2 \approx (2.5)/(0.5)$
$(2x)^2 \approx (2.5)/(0.5)(3.7\text{x}10^7)$
$4x^2 \approx 1.35\text{x}10^{-7}$
$x^2 \approx 1.35\text{x}10^{-7}/4 = 3.38\text{x}10^{-8}$
$x \approx \sqrt{3.38\text{x}10^{-8}} = 1.8\text{x}10^{-4}$
The assumption is valid

$[C] = 2.5 - x = 2.5 - 0.00018 = 2.5$ M

Believe it or not, the only reason we have to do it this way is, most calculators handle it better. (Try doing it the way we did with $K_c = 3.7\text{x}10^{-7}$, but using $K_c = 3.7\text{x}10^7$, and you'll see what happens: unless you keep all digits up to several decimal places, somebody's final concentration will be 0, which is impossible.)

If you have zero of one of the compounds to start with, the reaction has to go in the direction that produces it. But when you don't have any zeroes to start with, how do you know which direction it goes?

2 atm PCl_5, 0.005 atm PCl_3, and 0.005 atm Cl_2 are introduced into a closed container. What will be the pressure of each after the following reaction, whose K_p is 7.4×10^{-7}: $PCl_5 \Leftrightarrow PCl_3 + Cl_2$?

What we can do is, see what the ratio of products to reagents is right now, relative to where it wants to be. This is called the reaction quotient, Q, and has the same form as K.

$$Q_p = (P_{PCl3})(P_{Cl2})/(P_{PCl5}) = (0.005)(0.005)/2 = 1.25 \times 10^{-5}$$

$Q_p > K_p$, so we already have too much products. The reverse reaction will dominate, to lower products and increase reagents, until equilibrium is reached. By the same token, if Q_p were $< K_p$, it would mean we don't have enough products, so reaction would go forward. And if Q were = K, the system would already be at equilibrium and there would be no net reaction.

	$PCl_5 \Leftrightarrow$	$PCl_3 +$	Cl_2
I	2 atm	0.005 atm	0.005 atm
C	+x	-x	-x
E	2+x	0.005-x	0.005-x

$K_p = (P_{PCl3})(P_{Cl2})/(P_{PCl5})$
$7.4 \times 10^{-7} = (0.005-x)(0.005-x)/(2+x)$

Since every number is paired with an x, if we assume x is small all the x's will disappear and we won't be able to solve it! The equation is quadratic anyway, so we can solve it without assumptions.

$(2+x)(7.4 \times 10^{-7}) = (0.005-x)(0.005-x)$
$1.5 \times 10^{-6} + 7.4 \times 10^{-7}x = 2.5 \times 10^{-5} - 0.005x - 0.005x + x^2$
$0 = x^2 - 0.005x - 0.005x - 7.4 \times 10^{-7}x + 2.5 \times 10^{-5} - 1.5 \times 10^{-6}$
$0 = x^2 - 0.01x + 2.35 \times 10^{-5}$

$x = (-b \pm \sqrt{(b^2 - 4ac)})/2a$
$x = (+0.01 \pm \sqrt{((-0.01)^2 - 4(1)(2.35 \times 10^{-5})))}/(2)(1)$
$= (0.01 + 0.00245)/2$ or $(0.01 - 0.00245)/2$
$= 0.0622$ or 0.00378

The biggest x can be is 0.005, because otherwise you'd get a negative pressure for each product at equilibrium!
$x = 0.00378$

$P_{PCl_5} = 2 + x = 2.0$ atm
$P_{PCl_3} = 0.005 - x = 0.0012$ atm
P_{Cl_2} is also 0.0012 atm

Check this answer:
$K_p = (P_{PCl_3})(P_{Cl_2})/(P_{PCl_5}) = (0.0012)(0.0012)(2) = 7.2 \times 10^{-7}$ which is within rounding error of the given value of 7.4×10^{-7}.

Acids and bases

Only gases and solutions figure in equilibrium. The gas problems that are asked tend to be general, as shown above, whereas you are expected to know many specific kinds of solutions.

In the chapter on molecules, we saw that compounds with H^+ are usually acids and those with OH^- are usually bases. But this is not always true, and there are acids and bases without these features. Here are three common definitions.

	Acid	Base
Arrhenius	Makes H^+ in water	Makes OH^- in water
Brønsted-Lowry	Donates H^+	Accepts H^+
Lewis	Accepts e^-	Donates e^-

Lewis' definition is the most broadly correct, and is the basis of organic chemistry. For general chemistry, Arrhenius' is convenient.

Acids and bases can be ionic or covalent. And a solution can be anywhere in a whole spectrum of acidic to basic behaviour. The biggest $[H^+]$ or $[OH^-]$ you are likely to encounter in real life is around 20 M.

Nature	*acidic*	*neutral*	*basic*
pH	*-1*	*7*	*15*
[H⁺]	*10 M*	10^{-7} *M*	10^{-15} *M*
pOH	*15*	*7*	*-1*
[OH⁻]	10^{-15} *M*	10^{-7} *M*	*10 M*

"p" means "-log". It's a way to turn a number in scientific notation into something more handy.

pH = -log[H⁺]
pOH = -log[OH⁻]
pK = -logK

The numbers of the scale come from the dissociation (auto-hydrolysis) of water.

H$_2$O$_{(l)}$ ⇔	H$^+_{(aq)}$ +	OH$^-_{(aq)}$
I	-	0 mol/L	0 mol/L
C	-x	+x	+x
E	-	x	X

K$_c$ (called **K$_w$** for this reaction) **is 1.0x10^{-14}**

$K_w = [H^+][OH^-]$
$1.0 \times 10^{-14} = (x)(x)$
$1.0 \times 10^{-14} = x^2$
$\sqrt{(1.0 \times 10^{-14})} = x$
$1.0 \times 10^{-7} = x$

$[H^+] = x = 1.0 \times 10^{-7}$, pH = 7
$[OH^-] = x = 1.0 \times 10^{-7}$, pOH = 7

That's for neutral water. If you add a substance to the water, the water can become acidic or basic, but the $[H^+][OH^-]$ has to remain = 1.0×10^{-14}. In other words, **pH + pOH = 14** for any solution. H$^+$ is sometimes shown attached to H$_2$O, making H$_3$O$^+$ but behaving the same way.

Strong acid
Ionic compound that dissolves 100% in water, to give H$^+$ and an inert (spectator) ion. HCl, HBr, HI, HNO$_3$, HClO$_4$, and H$_2$SO$_4$ are common. If you're in school, your teacher may give you a list to memorize.

Find the pH of a 0.65 M solution of HCl.

HCl$_{(aq)}$ →	H$^+_{(aq)}$ +	Cl$^-_{(aq)}$
I	0.65 M	0 M	0 M
C	-0.65	+0.65	+0.65
F	0	0.65	0.65

[H+] = 0.65 M
pH = -log(0.65) = 0.19

Strong base
Ionic compound that dissolves 100% in water, to give OH$^-$ and a spectator. Typically, a column I element with OH: LiOH, NaOH, etc.

Find the pH of a 0.65 M solution of KOH.

	KOH$_{(aq)}$ →	K$^+_{(aq)}$ +	OH$^-_{(aq)}$
I	0.65 M	0 M	0 M
C	-0.65	+0.65	+0.65
F	0	0.65	0.65

$[OH^-] = 0.65$ M
$pOH = -\log(0.65) = 0.2$
$pH + pOH = 14$
$pH = 14 - pOH = 13.8 = 14$ (to 2 sig figs)

Weak acid

Covalent compound that reacts <100% with water to make H_3O^+ and a "conjugate" base ion. All acids other than the strong ones are weak: HF, HCH_3CO_2 ("acetic"), $HCHO_2$ ("formic"), and so on.

Find the pH of a 0.65 M solution of HF, given $K_a = 6.3x10^{-4}$.

	HF$_{(aq)}$ +	H$_2$O$_{(l)}$ ⇔	H$_3$O$^+_{(aq)}$ +	F$^-_{(aq)}$
I	0.65 M	-	0 M	0 M
C	-x	-	+x	+x
E	0.65-x	-	x	X

K_c (called **K_a** for weak acids) = $[H_3O^+][F^-]/[HF]$
$6.3x10^{-4} = (x)(x)/(0.65-x)$

Assume x < (5%)(0.65) = 0.0325
$6.3x10^{-4} \approx x^2/0.65$
$0.0202 \approx x$
The assumption is valid

$[H_3O^+] = x = 0.0202$ M
$pH = -\log[H_3O^+] = 1.7$

What is the percent dissociation?

% dissociation = (amount dissolved/initial amount)(100%)
= (x/0.65)(100%) = (0.0202/0.65)(100%) = 3.1%

Weak base

Covalent compound that reacts <100% with water to make OH$^-$ and a conjugate acid. Includes all non-strong bases: NH_3 ("ammonia"), NH_2CH_3 ("methylamine"), and so on.

Find the pH of a 0.65 M solution of NH_3, given $K_b = 1.8x10^{-5}$.

	$NH_{3(aq)} +$	$H_2O_{(l)} \Leftrightarrow$	$OH^-_{(aq)} +$	$NH_4^+_{(aq)}$
I	0.65 M	-	0 M	0 M
C	-x	-	+x	+x
E	0.65-x	-	x	x

K_c (called K_b now) $= [OH^-][NH_4^+]/[NH_3]$
$1.8\times10^{-5} = (x)(x)/(0.65-x)$

Assume $x < (5\%)(0.65) = 0.0325$
$1.8\times10^{-5} \approx x^2/0.65$
$0.00342 \approx x$
The assumption is valid

$[OH^-] = x = 0.00342$ M
$pOH = 2.5$
$pH = 11.5 = 12$ (to 2 sig figs)

You can have a solution of just the conjugate. Sometimes this will be attached to a spectator ion, which you can ignore.

Find the pH of a 0.65 M solution of NH_4NO_3, given K_b of NH_3 is 1.8×10^{-5}.

NH_4^+ is the conjugate acid of NH_3
NO_3^- is a spectator

	$NH_4^+ +$	$H_2O \Leftrightarrow$	$H_3O^+ +$	NH_3
I	0.65 M	-	0 M	0 M
C	-x	-	+x	+x
E	0.65-x	-	X	x

$K_a = [H_3O^+][NH_3]/[NH_4^+]$
But where is its value?
Get it from K_b.
Since $[H^+][OH^-] = K_w$, $(K_a)(K_b) = K_w$, for conjugates.
$(K_a$ of $NH_4^+)(K_b$ of $NH_3) = K_w$
$K_a = K_w/K_b = 1.0\times10^{-14}/1.8\times10^{-5} = 5.56\times10^{-10}$

$5.56\times10^{-10} = (x)(x)/(0.65-x)$

Assume $x < (5\%)(0.65) = 0.0325$
$5.56\times10^{-10} \approx x^2/0.65$
$1.90\times10^{-5} \approx x$
The assumption is valid

$[H_3O^+] = x = 1.90 \times 10^{-5}$
pH = 4.7

You can have a mixture of acids or bases. Usually, it will be two strong ones or one strong and one weak, because the math is more complicated with two weak ones. The rule of thumb is, do strong before weak.

Find the pH of a solution of 0.65 M NH_3 ($K_b = 1.8 \times 10^{-5}$) in 0.025 M NaOH.

NaOH is strong base
NH_3 is weak base

	NaOH →	Na+ +	OH-
I	0.025 M	0 M	0 M
C	-0.025	+0.025	+0.025
F	0	0.025	0.025

	NH_3 +	H_2O ⇔	OH- +	NH_4^+
I	0.65 M	-	0.025 M	0 M
C	-x	-	+x	+x
E	0.65-x	-	0.025+x	x

Notice the initial [OH-] is not 0, because we have some from an outside source (common ion).

$K_b = [OH^-][NH_4^+]/[NH_3]$
$1.8 \times 10^{-5} = (0.025+x)(x)/(0.65-x)$

Assume x < (5%)(0.025) = 0.00125
Automatically, then, x ≪ 0.65
$1.8 \times 10^{-5} \approx (0.025)(x)/0.65$
$0.000468 \approx x$
The assumption is valid

$[OH^-] = 0.0025 + x = 0.0030$
pOH = 2.5
pH = 12

Some acids ("polyprotics") and bases can produce multiple H+'s or OH-'s.

Find the pH of a 0.65 M solution of H_2CO_3, given that $K_{a1} = 2.5 \times 10^{-4}$ and $K_{a2} = 5.6 \times 10^{-11}$.

	H_2CO_3 +	H_2O ⇔	H_3O^+ +	HCO_3^-
I	0.65 M	-	0 M	0 M
C	-x	-	+x	+x
E	0.65-x	-	x	x

	HCO_3^- +	H_2O ⇔	H_3O^+ +	CO_3^{-2}
I	x	-	x M	0 M
C	-y	-	+y	+y
E	x-y	-	x+y	y

$K_{a1} = [H_3O^+][HCO_3^-]/[H_2CO_3]$
$2.5 \times 10^{-4} = (x+y)(x-y)/(0.65-x)$

$K_{a2} = [H_3O^+][CO_3^{-2}]/[HCO_3^-]$
$5.6 \times 10^{-11} = (x+y)(y)/(x-y)$

Notice that you have to wait until the end to get the true concentrations of each species. For example, $[H_3O^+]$ is now (x+y), no longer just x.

This gives a system of equations, which is harder to solve than the one we did earlier. However, here too a simplifying assumption will usually work.

Since $K_{a1} \gg K_{a2}$, x ≫ y

K_{a1}
$2.5 \times 10^{-4} \approx (x)(x)/(0.65-x)$

Further assume that x < (5%)(0.65) = 0.0325
$2.5 \times 10^{-4} \approx (x)(x)/(0.65)$
$0.0128 \approx x$

K_{a2}
$5.6 \times 10^{-11} \approx (x)(y)/(x)$
$5.6 \times 10^{-11} \approx y$

All assumptions are valid

$[H_3O^+] = x + y = 0.0128 + 5.6 \times 10^{-11} = 0.0128$
pH = 1.9

As with a mixture of strong and weak acid (or strong and weak base), the $[H_3O^+]$ or $[OH^-]$ will usually be dominated by the first reaction.

If you have a multi-step conjugate, be careful which K_a you relate to which K_b. For example, in a solution of CO_3^{-2}:

$CO_3 + H_2O \Leftrightarrow OH^- + HCO_3^-$

$K_{b1} = [OH^-][HCO_3^-]/[CO_3^{-2}] = K_w/K_{a2}$, not K_w/K_{a1} (since K_{a1} doesn't contain CO_3^{-2})

$HCO_3^- + H_2O \Leftrightarrow OH^- + H_2CO_3$

$K_{b2} = [OH^-][H_2CO_3]/[HCO_3^-] = K_w/K_{a1}$, not K_w/K_{a2} (since K_{a2} doesn't contain H_2CO_3)

Also notice that here HCO_3^- is a base, whereas in the previous example it was an acid. Some substances can do this. They're "amphoteric".

Buffers

A solution of a weak acid or base and its conjugate has a special property: it can neutralize incoming acid or base, in a way that keeps pH (and pOH) relatively stable. Acid-base reaction is 1-way, no matter whether strong or weak.

$NH_3 + H^+ \rightarrow NH_4^+$
$NH_4^+ + OH^- \rightarrow NH_3 + H_2O$

This is a buffer, and it keeps you alive. Your saliva, blood, and stomach acid, for example, all have enzymes (catalysts) that only function over a certain pH range. Imagine if eating an orange suddenly made your whole body so acidic it stopped working properly!

There is a catch: if you throw so much acid at the buffer, eventually you deplete the base and you no longer have a buffer. Likewise for dumping base on it. You can survive swallowing a little soap, which is very basic, but don't go eating whole bars of it. So, in general, we say that the buffer works when the proportion of acid half to base half is between 1/10 and 10/1.

Which of the following solutions are buffers?
a) 0.5 M HCl with 0.5 M NaCl
b) 0.5 M HF with 0.4 M NaF
c) 0.5 M H_2S with 0.5 M K_2S
d) 0.5 M NH_3 with 0.03 M NH_4NO_3

a) HCl is strong acid, so no buffer.
b) HF is weak acid, F^- is its conjugate (Na^+ is a spectator), so this is a buffer.
c) H_2S is weak, but S^{-2} is not its direct conjugate (HS^- is). No buffer.
d) NH_3 and NH_4^+ are conjugates, but the proportion (0.5/0.03) is greater than 10/1, so no buffer.

What is the pH of a solution of 0.5 M HF ($K_a = 6.3x10^{-4}$) and 0.4 M NaF?

Set it up using the reaction corresponding to the K given.

	HF +	H_2O ⇔	H_3O^+ +	F^-
I	0.5 M	-	0	0.4
C	-x	-	+x	+x
E	0.5-x	-	x	0.4+x

$K_a = [H_3O^+][F^-]/[HF]$
$6.3 \times 10^{-4} = (x)(0.4+x)/(0.5-x)$

Assume $x < (5\%)(0.4) = 0.02$
$6.3 \times 10^{-4} \approx (x)(0.4)/0.5$
$7.88 \times 10^{-4} \approx x$
The assumption is valid

$[H_3O^+] = x$
$pH = 3.1$

With buffers, there is a convenient shortcut for the math, known as the Henderson-Hasselbach equations.

Since $K_a = [H_3O^+][\text{conjugate base}]/[\text{acid}]$
$-\log K_a = -\log([H_3O^+][\text{conjugate base}]/[\text{acid}])$
$-\log K_a = -\log[H_3O^+] - \log([\text{conjugate base}]/[\text{acid}])$
$pK_a = pH - \log([\text{conjugate base}]/[\text{acid}])$
$pK_a + \log([\text{conjugate base}]/[\text{acid}]) = pH$

Likewise,
$pK_b + \log([\text{conjugate acid}]/[\text{base}]) = pOH$

Since the buffer works when [conjugate acid or base]/[base or acid] is between 1/10 and 10/1, and since log 1/10 = -1 and log 10/1 = +1, we say that the buffer's effective pH or pOH range is pK_a or pK_b ±1.

What happens to the pH if you add 10 mL of 0.6 M NaOH to 100 mL of the above buffer?

The OH⁻ (strong base) will react with the buffer's acid part, HF.
$HF + OH^- \rightarrow H_2O + F^-$

Before filling in the I-C-F grid of calculations, any time you mix two solutions, you must account for the effect of dilution.

NaOH: (0.6 mol/L)(0.010 L) = 0.006 mol
This gives 0.006 mol OH⁻
The total volume is 110 mL, or 0.11 L
So now [OH⁻] = 0.006 mol/0.11 L = 0.054 M

HF: $(0.5 \text{ M})(0.1 \text{ L}) = 0.05$ mol
0.05 mol$/0.11$ L $= 0.454$ M

F⁻: $(0.4 \text{ M})(0.1 \text{ L}) = 0.04$ mol
0.04 mol$/0.11$ L $= 0.364$ M

	HF +	OH⁻ →	H₂O +	F⁻
I	0.454 M	0.054	-	0.364
C	-0.054	-0.054	-	+0.054
F	0.400	0	-	0.418

We still have a buffer.
pH = pKₐ + log([conjugate base]/[acid])
= -log(6.3×10^{-4}) + log(0.418 M/0.400 M) = 3.22 = 3 (to 1 sig fig)

Titration

Titration is the process of adding acid to base or base to acid, slowly enough so that you can detect when the neutralization (equivalence point, E.P.) is complete. Often, you add a few drops of a compound (indicator) that changes colour when you get close to the E.P.

Find the pH after adding 0, 8, 10, and 11 mL of 0.5 M HNO_3 to 20 mL of 0.25 M NH_2CH_3 ($K_b = 4.4 \times 10^{-4}$).

a) After 0 mL, we have only NH_2CH_3 weak base.

	NH₂CH₃ +	H₂O ⇔	OH⁻ +	NH₃CH₃⁺
I	0.25 M	-	0	0
C	-x	-	+x	+x
E	0.25-x	-	x	X

$K_b = [OH^-][NH_3CH_3^+]/[NH_2CH_3]$
$4.4 \times 10^{-4} = (x)(x)/(0.25-x)$

Assume x < (5%)(0.25) = 0.0125
$4.4 \times 10^{-4} \approx (x)(x)/0.025$
$0.0105 \approx x$
The assumption is valid

$[OH^-] = x$
pOH = 2.0
pH = 12

b) After 8 mL, we have NH_2CH_3 weak base and H^+ strong acid.

Diluted $[NH_2CH_3]$ = (0.25 M)(0.02 L)/0.028 L = 0.179 M
Diluted $[H^+]$ = (0.5 M)(0.008 L)/0.028 L = 0.143 M

	NH_2CH_3 +	$H^+ \rightarrow$	$NH_3CH_3^+$
I	0.179	0.143	0
C	-0.143	-0.143	+0.143
F	0.036	0	0.143

Now we have a buffer.
pOH = pK_b + log([conjugate acid]/[base])
= -log($4.4x10^{-4}$) + log(0.143/0.036) = 4.0
pH = 10
It makes sense that when we added acid, the pH went down.

c) After 10 mL:

$[NH_2CH_3]$ = (0.25 M)(0.02 L)/0.03 L = 0.167 M
$[H^+]$ = (0.5 M)(0.01 L)/0.03 L = 0.167 M

	NH_2CH_3 +	$H^+ \rightarrow$	$NH_3CH_3^+$
I	0.167 M	0.167	0
C	-0.167	-0.167	+0.167
F	0	0	0.167

This is the E.P.!
Now we have only $NH_3CH_3^+$ conjugate acid.

	$NH_3CH_3^+$ +	$H_2O \Leftrightarrow$	H_3O^+ +	NH_2CH_3
I	0.167 M	-	0	0
C	-x	-	+x	+x
E	0.167-x	-	x	x

$K_a = K_w/K_b$ = ($1.0x10^{-14}/4.4x10^{-4}$) = $2.27x10^{-11}$

$K_a = [H_3O^+][NH_2CH_3]/[NH_3CH_3^+]$
$2.27x10^{-11}$ = (x)(x)/(0.167-x)

Assume x < (5%)(0.167) = 0.0083
$2.27x10^{-11} \approx$ (x)(x)/0.167
$1.95x10^{-6} \approx$ x
The assumption is valid

$[H_3O^+]$ = 1.95x10^{-6} M
pH = 5.7
Good, the pH is still dropping.

d) After 11 mL:

$[NH_2CH_3]$ = (0.25 M)(0.02 L)/0.031 L = 0.161 M
$[H^+]$ = (0.5 M)(0.011 L)/0.031 L = 0.177 M

	NH_2CH_3 +	H^+ →	$NH_3CH_3^+$
I	0.161	0.177	0
C	-0.161	-0.161	+0.161
F	0	0.016	0.161

Now we have a mixture of acids, $NH_3CH_3^+$ and H^+.

	$NH_3CH_3^+$ +	H_2O ⇌	H_3O^+ +	NH_2CH_3
I	0.161	-	0.016	0
C	-x	-	+x	+x
E	0.161-x	-	0.016+x	x

$K_a = [H_3O^+][NH_2CH_3]/[NH_3CH_3^+]$
$2.27\text{x}10^{-11}$ = (0.016+x)(x)/(0.161-x)

Assume x < (5%)(0.016) = 0.00080
$2.27\text{x}10^{-11} \approx$ (0.016)(x)/0.161
$2.29\text{x}10^{-10} \approx$ x
The assumption is valid

$[H_3O^+]$ = 0.016 + x = 0.016 + 2.29x10^{-10} = 0.016 M
pH = 1.8

The progression is shown by a titration curve.

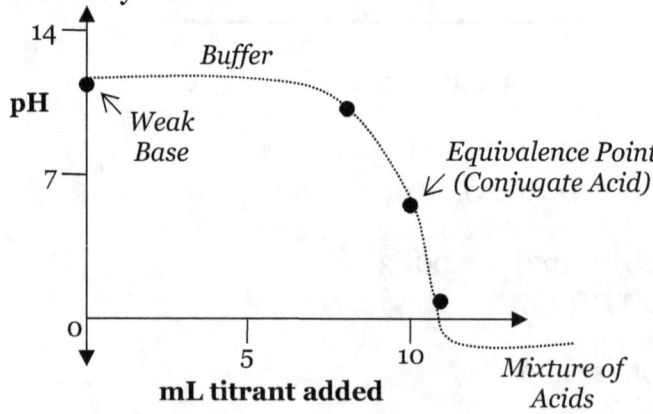

Typically, four kinds of titrations are done. In the curves below, notice the pH at the equivalence point.

Titration	Products	Example	Curve
Add strong acid to weak base	Salt of conjugate (acid)	$HCl + NH_3 \rightarrow$ NH_4Cl	
Add strong acid to strong base	Neutral salt plus water	$HCl + NaOH \rightarrow$ $NaCl + H_2O$	
Add strong base to weak acid	Salt of conjugate (base) plus water	$NaOH + HF \rightarrow$ $NaF + H_2O$	
Add strong base to strong acid	Neutral salt plus water	$NaOH + HCl \rightarrow$ $NaCl + H_2O$	

Indicators

There are three main kinds of indicators for this course.

Two-colour indicators change colour over a small pH range. They are actually just buffers, whose acid half is one colour and base half is another. For example, bromocresol green has an acid, yellow form (symbolized as HIn), and a conjugate base, blue form (In-). Its K (called K_{In}) is 2.09×10^{-5}. Its effective pH range is estimated to be $pK_{In} \pm 1$, or 3.7 to 5.7. (In real life, it turns out to be 4.0 to 5.6). At pH's below 4.0, the acid form dominates, and the environing solution is yellow. At pH's above 5.6, the base form dominates and the solution is blue. In between, the colour is changing.

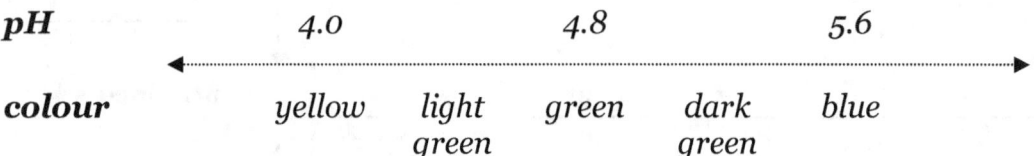

pH	4.0		4.8		5.6
colour	yellow	light green	green	dark green	blue

Litmus paper is essentially a two-colour indicator with $pK_{In} = 7$. It is red in acid and blue in base.

Universal indicators combine several different two-colour indicators, each with a different pH range and set of colours. In this way, no matter where your solution is in the pH spectrum, one of the indicators is working. For example:

Component	pH range	Colours
Thymol blue	1.2-2.8	red-yellow
Methyl red	4.4-6.2	red-yellow
Bromothymol blue	6.0-7.6	yellow-blue
Thymol blue	8.0-9.6	yellow-blue
Phenolphthalein	8.2-10	clear-purple

pH	0	2	4	6	8	10	12
colour	red	orange	yellow	green		blue	

Salts

Broadly speaking, any ionic compound is a salt. More narrowly, a salt is understood to be any neutral ionic compound other than an acid or base.

Soluble salt

Ionic compound that dissolves 100% in water, based on solubility rules.

Find the concentration of all ions in a 0.65 M solution of $CaCl_2$.

	CaCl$_{2(aq)}$	→ Ca$^{+2}_{(aq)}$ +	2Cl$^-_{(aq)}$
I	0.65 M	0 M	0 M
C	-0.65	+0.65	+2(0.65)
F	0	0.65	1.30

Insoluble salt

Ionic compound that dissolves <100% in water. All salts other than the soluble ones are insoluble.

Find the solubility and concentration of all ions in a solution of PbCl$_2$ (K$_{sp}$ = 1.6x10^{-5}).

	PbCl$_{2(s)}$ ⇌	Pb$^{+2}_{(aq)}$ +	2Cl$^-_{(aq)}$
I	-	0 M	0 M
C	-x	+x	+2x
E	-	x	2x

K$_c$ (called **K$_{sp}$** for insoluble salts) = [Pb^{+2}][Cl$^-$]^2
1.6x10^{-5} = (x)(2x)^2
1.6x10^{-5} = (x)(4x^2)
1.6x10^{-5} = 4x^3
$^3\sqrt{(1.6x10^{-5}/4)}$ = x
0.0159 = x

[Pb^{+2}] = x = 0.016 M
[Cl$^-$] = 2x = 2(0.0159) = 0.032 M
"Solubility" = amount dissolved = x = 0.016 M

As with acids and bases, you can have a mixture of salts, and to solve it do the 1-way reaction first.

What is the solubility of PbCl$_2$ in 0.7 M NaCl?

NaCl is a soluble salt
PbCl$_2$ is an insoluble salt

	NaCl$_{(aq)}$ →	Na$^+_{(aq)}$ +	Cl$^-_{(aq)}$
I	0.7 M	0 M	0 M
C	-0.7	+0.7	+0.7
F	0	0.7	0.7

$PbCl_{2(s)}$ $\Leftrightarrow$	$Pb^{+2}_{(aq)}$ +	$2Cl^{-}_{(aq)}$
I	-	0 M	0.7 M
C	-x	+x	+2x
E	-	X	0.7+2x

$K_{sp} = [Pb^{+2}][Cl^-]^2$
$1.6x10^{-5} = (x)(0.7+2x)^2$

Assume $2x < (5\%)(0.7) = 0.035$
$1.6x10^{-5} \approx (x)(0.7)^2$
$3.3x10^{-5} \approx x$
The assumption is valid.

Solubility = x = $3.3x10^{-5}$ M

Notice that this makes sense according to Le Chatelier: it's as if we added some Cl- to an equilibrium of $PbCl_2$; the system wants to use up the added Cl-, and does this by going backward; so less $PbCl_2$ dissolves.

By now, you should always be asking yourself if your answers make real-life sense, and fixing them if they don't. Don't let a recopying or calculator error make you report, say, a pH of 13 for an acid!

A solution contains copper(I) and tin(II) ions, at a concentration of 0.0005 M each, and your task is to separate them. You do this by slowly adding sulfide ion. One of the metals will precipitate first. However, the separation only works if 99.99% of the first ion is precipitated before the second one starts to. Does it work in this case? K_{sp} of copper(I)sulfide is $1.1x10^{-47}$ and K_{sp} of tin(II)sulfide is $7.0x10^{-26}$.

The reactions corresponding to the K_{sp}'s are:

$Cu_2S_{(s)} \Leftrightarrow 2Cu^+_{(aq)} + S^{-2}_{(aq)}$
$K_{sp} = [Cu^+]^2[S^{-2}]$

$SnS_{(s)} \Leftrightarrow Sn^{+2}_{(aq)} + S^{-2}_{(aq)}$
$K_{sp} = [Sn^{+2}][S^{-2}]$

In each case, we can see what $[S^{-2}]$ would be at equilibrium.

$K_{sp} = [Cu^+]^2[S^{-2}]$
$1.1x10^{-47} = [0.0005 M]^2[S^{-2}]$
$4.4x10^{-41} = [S^{-2}]$

$K_{sp} = [Sn^{+2}][S^{-2}]$
$7.0x10^{-26} = [0.0005 M][S^{-2}]$

$1.4 \times 10^{-22} = [S^{-2}]$

If you add a drop more S^{-2}, the system will want to use it up and make a precipitate. It takes less S^{-2} to get the Cu^+ started precipitating, so Cu_2S precipitates first.

When 99.99% of the Cu^+ is gone, 0.01% is left, so the concentration is:
$(0.01\%)(0.0005 \text{ M}) = (0.0001)(0.0005) = 5 \times 10^{-8} \text{ M}$

Now, the concentration of S^{-2} is:
$K_{sp} = [Cu^+]^2[S^{-2}]$
$1.1 \times 10^{-47} = [5 \times 10^{-8} \text{ M}]^2[S^{-2}]$
$4.4 \times 10^{-33} = [S^{-2}]$

It's still too low to start precipitating the Sn^{+2} ion, so the separation is successful.

Complexes

Transition metal complexes, discussed in the chapter on molecules, are also governed by reversible reactions.

What is the concentration of all ions in the dissociation of 1.0 M $Ni(NH_3)_6^{+2}$, given K_h is 1.8×10^{-9}?

	$Ni(NH_3)_6^{+2}$ (aq) $\Leftrightarrow$	Ni^{+2}(aq) +	$6NH_3$ (aq)
I	1.0 M	0 M	0 M
C	-x	+x	+6x
E	1.0-x	x	6x

K_c (called $\mathbf{K_h}$ for complexes) $= [Ni^{+2}][NH_3]^6/[Ni(NH_3)_6^{+2}]$
$1.8 \times 10^{-9} = (x)(6x)^6/(1.0-x)$

Assume $x < (5\%)(1.0) = 0.05$
$1.8 \times 10^{-9} \approx (x)(6x)^6/1.0$
$1.8 \times 10^{-9} \approx (x)(46656x^6)$
$3.86 \times 10^{-14} \approx x^7$
$0.0121 \approx x$
The assumption is valid

The ion concentrations are:
$[Ni^{+2}] = x = 0.012 \text{ M}$
$[NH_3] = 6x = 0.073 \text{ M}$

Thermodynamics:
Does a Process Want to Happen?

Enthalpy

When we looked at balancing chemical equations, we saw that matter has to be conserved. It turns out that energy does too. Thermodynamics is the study of where energy comes from and where it goes.

In general, breaking bonds requires "enthalpic" energy (often called heat, which as we'll see is slightly different) and forming them releases it. Every bond is different. Here are some sample bond energies, in kJ/mol:

H-H	435		C-O	360
C-H	414		C=O	798
C-C	347		H-O	464
C=C	611		O-O	142
C≡C	837		O=O	498

The enthalpy change (ΔH) of a process can be found by totaling all the bonds broken and formed. Any bond energies you need to solve a problem will be given in the problem statement or in an attached table of data.

How much enthalpy is released or absorbed in the combustion of C_3H_8?

We need a balanced chemical reaction:

$$C_3H_8 + \qquad\qquad 5O_2 \rightarrow \quad 3CO_2 + \qquad\qquad 4H_2O$$

And we need Lewis structures, to see the bonds:

$$:\ddot{O}=\ddot{O}:$$
$$:\ddot{O}=\ddot{O}:$$

H H H $\qquad\qquad :\ddot{O}=\ddot{O}:\qquad\qquad\qquad$ H–$\ddot{O}$–H
| | |
H– C– C– C–H $\qquad :\ddot{O}=\ddot{O}:\qquad :\ddot{O}=C=\ddot{O}:\qquad$ H–$\ddot{O}$–H
| | | $\qquad\qquad :\ddot{O}=\ddot{O}:\qquad :\ddot{O}=C=\ddot{O}:\qquad$ H–$\ddot{O}$–H
H H H $\qquad\qquad :\ddot{O}=\ddot{O}:\qquad :\ddot{O}=C=\ddot{O}:\qquad$ H–$\ddot{O}$–H

Reagents	Products
(bonds broken)	(bonds formed)
2 C–C = 2(347) = 694	6 C=O = 6(798) = 4788
8 C–H = 8(414) = 3312	8 H–O = 8(464) = 3712
5 O=O = 5(498) = 2490	
Total = 6496	Total = 8500

Holy Holmium! Complete General Chemistry in 150 Pages

Net enthalpy released = 8500 - 6496 = 2004 kJ/mol

The experimental value turns out to be 2220 kJ/mol. The problem with the calculation is, bond energies are averages taken by looking at many molecules. In real life, for example, a C-C bond in C_3H_8 is not worth the same as a C-C bond in another compound.

By convention, the release of enthalpy is "exothermic" and has a negative ΔH, while the absorption of enthalpy is "endothermic" and has a positive ΔH. In the above example, then, ΔH = -2004 kJ/mol. In general, processes like to be exothermic: they release pent-up energy and wind up more stable. Imagine a bomb, or someone on the verge of a temper tantrum.

In a similar way, the enthalpy change of a process is the difference between the stabilities of products and reagents, as measured by the standard enthalpy of formation, (ΔH^o_f). ΔH^o_f reflects how hard or easy it is to make a compound from its elements at standard temperature and pressure. By definition, it takes no energy to make an element in its standard state. Here are some sample values of ΔH^o_f, in kJ/mol:

$O_{2(g)}$	0	$C_2H_{4(g)}$	+52
$O_{2(l)}$	-7	$C_2H_{2(g)}$	+227
$CH_{4(g)}$	-75	$CO_{2(g)}$	-394
$C_2H_{6(g)}$	-85	$H_2O_{(g)}$	-242
$C_3H_{8(g)}$	-104	$H_2O_{(l)}$	-286

Any ΔH^o_f's you need to solve a problem will be given in one way or another.

Give the reaction corresponding to the formation of CH_4.

Carbon in standard state + Hydrogen in standard state → CH_4
$C(s) + 2H_2(g) \rightarrow CH_4(g)$

ΔH^o = products − reagents
= $\Delta H^o_{CH4} - (\Delta H^o_C + 2\Delta H^o_{H2})$
= -75 - [0 + 2(0)] = -75 kJ/mol

Find the enthalpy of combustion of C_3H_8.

$C_3H_{8(g)} + 5O_{2(g)} \rightarrow 3CO_{2(g)} + 4H_2O_{(l)}$

ΔH^o = products − reagents
= $(3\Delta H^o_{CO2} + 4\Delta H^o_{H2O}) - (\Delta H^o_{C3H8} + 5\Delta H^o_{O2})$
= [3(-394) + 4(-286)] − [(104) + 5(0)]
= -2222 kJ/mol

Notice that you get a much more accurate answer with ΔH°_f's than with bond energies.

A few things follow from all of this.

- If you reverse a reaction, reverse the sign on ΔH
- If you multiply a reaction, multiply ΔH
- If you add one reaction to another, to make some bigger overall process, add their ΔH's.

That is Heinrich Hess' Law.

What is the enthalpy of sublimation of magnesium, given the following data, in kJ/mol?

Lattice energy of $MgCl_2$:	*2527*
Standard enthalpy of formation of $MgCl_2$:	*-642*
Cl-Cl bond energy:	*244*
1st ionization energy of Mg:	*738*
2nd ionization energy of Mg:	*1451*
Electron affinity of Cl:	*-349*

Write a reaction for each process.

A $Mg^{+2}_{(g)} + 2Cl^-_{(g)} \rightarrow MgCl_{2(s)}$ $\Delta H = -2527$
B $Mg_{(s)} + Cl_{2(g)} \rightarrow MgCl_{2(s)}$ $\Delta H = -642$
C $Cl_{2(g)} \rightarrow 2Cl_{(g)}$ $\Delta H = 244$
D $Mg_{(g)} \rightarrow Mg^+_{(g)} + e^-$ $\Delta H = 738$
E $Mg^+_{(g)} \rightarrow Mg^{+2}_{(g)} + e^-$ $\Delta H = 1451$
F $Cl_{(g)} + e^- \rightarrow Cl^-_{(g)}$ $\Delta H = -349$
G $Mg_{(s)} \rightarrow Mg_{(g)}$ $\Delta H = ?$

Figure out how to manipulate A-F to make G. One way to do this is, look at what compounds (and in what state) you need to have on the left and right sides of the equation, so that all of them cancel except the ones you want in the target equation.

Need $Mg_{(s)}$ on left, so keep B as it is:
$Mg_{(s)} + Cl_{2(g)} \rightarrow MgCl_{2(s)}$ $\Delta H = -642$

Need $Mg_{(g)}$ on right, so flip D:
$Mg^+_{(g)} + e^- \rightarrow Mg_{(g)}$ $\Delta H = -738$

Need $Mg^+_{(g)}$ on right to cancel with D, so flip E:
$Mg^{+2}_{(g)} + e^- \rightarrow Mg^+_{(g)}$ $\Delta H = -1451$

Need $Cl_{2(g)}$ on right to cancel with B, so flip C:
$2Cl_{(g)} \rightarrow Cl_{2(g)}$ $\Delta H = -244$

Need $MgCl_{2(s)}$ on left to cancel with B, so flip A:

$MgCl_{2(s)} \rightarrow Mg^{+2}_{(g)} + 2Cl^-_{(g)}$ $\Delta H = 2527$

Need $2Cl_{(g)}$ on right to cancel with C, so flip F and multiply by 2:

$2Cl^-_{(g)} \rightarrow 2Cl_{(g)} + 2e^-$ $\Delta H = 698$

Add them up:

$Mg_{(g)} \rightarrow Mg_{(s)}$ $\Delta H = 150$

The experimental value for $\Delta H_{sublimation}$ for Mg is 146.

This question lets you see how different tools learned in previous chapters fit together, and how to patiently unravel a question that may seem long or tough.

This way of doing the calculations is called a Born-Haber cycle. The step-by-step change can be shown in an energy diagram:

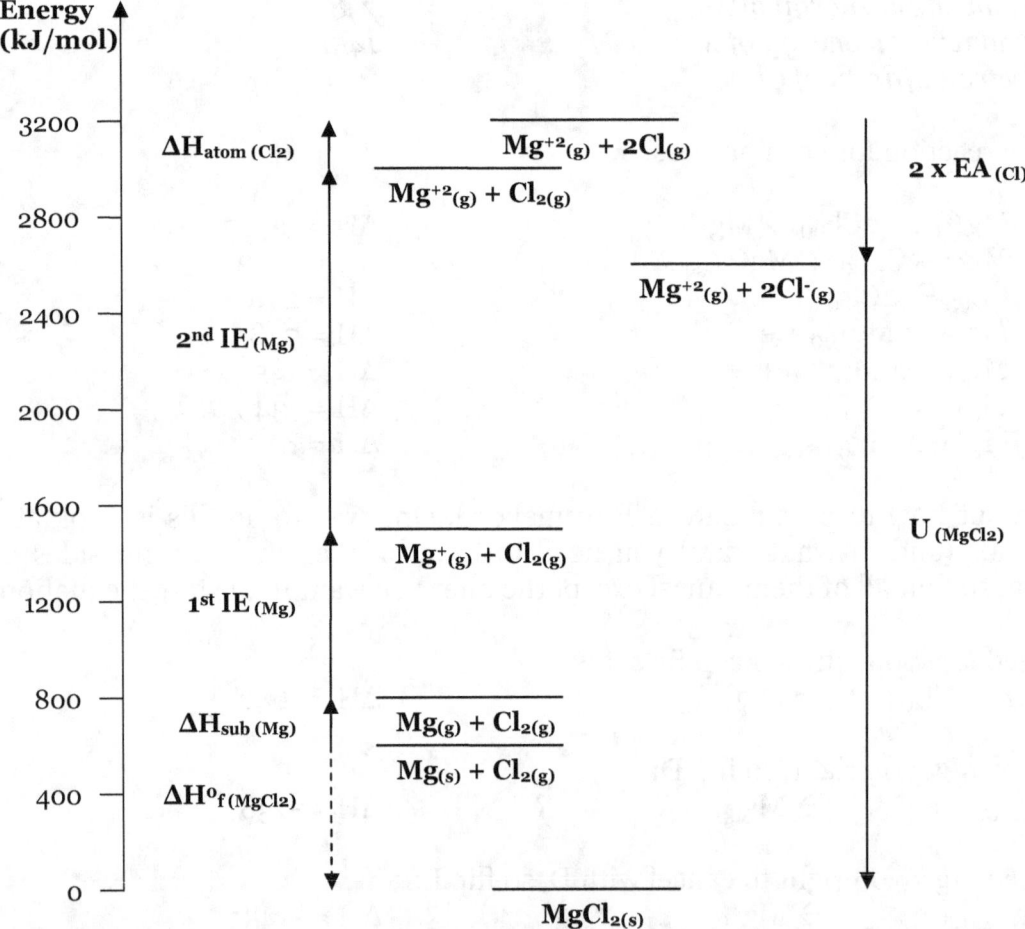

So can the overall change, using a different kind of energy diagram:

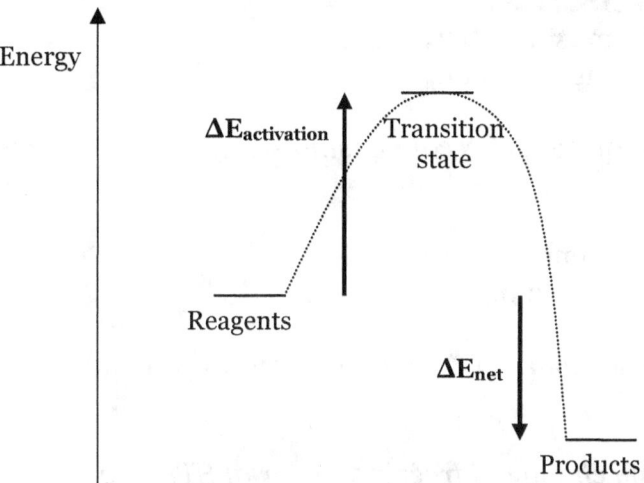

The transition state (also called "activated complex") is an imaginary half-way point in bond breakage and formation. For example, for the electrolysis of water, on a molecular level, it would look like this:

Reagents ($2H_2O$)	→	**Transition state**	→	**Products** ($2H_2 + O_2$)

Products minus reagents is the net internal energy change (ΔE_{net}) of reaction. Sometimes it's shown as ΔH or ΔG. ΔE and ΔG will both be presented in a moment. Meanwhile, transition state minus reagents gives the internal energy change of activation ($\Delta E_{activation}$).

Entropy

Whether or not a reaction wants to happen also has to do with molecules' freedom of motion (entropy, S). According to the Third Law of Thermodynamics, zero freedom is a solid at the lowest temperature theoretically possible: absolute zero, 0 K or -273 °C. The coldest temperature ever measured on Earth (in Antarctica) is -89 °C. But, humans have been able to create laboratory conditions close to absolute zero.

What follows from this is, as you raise the temperature, entropy goes up. And we saw with phase diagrams that this tends to mean going from solid to liquid to gas.

The Second Law of Thermodynamics says, things tend to evolve toward greater freedom. For example, astronomers have observed that the universe is expanding. In other words, for any process, $\Delta S_{universe} > 0$. Taking the universe as an arbitrarily defined system and its surroundings, therefore:

$$\Delta S_{univ} = \Delta S_{syst} + \Delta S_{surr} > 0$$
if ΔS_{syst} is +, ΔS_{surr} must be + or a small −
if ΔS_{syst} is −, ΔS_{surr} must be a big +

Predict the sign of ΔS_{surr} for condensing steam.

$$H_2O_{(g)} \Leftrightarrow H_2O_{(l)}$$
ΔS_{syst} is −, because entropy is lost
So ΔS_{surr} must be + and big

As with ΔH, reference data for ΔS will be given to you and can be manipulated using Hess' Law.

Find the entropy change of freezing water at STP, given:
$S^o_{H2O(g)} = 189 \ J/mol \cdot K$
$S^o_{H2O(l)} = 110 \ J/mol \cdot K$

ΔS^o = products − reagents = 110 - 189 = -79 J/mol·K
This fits our prediction.

What is the entropy change of the surroundings, for this process?

$\Delta S^o_{surr} + \Delta S^o_{syst} > 0$
$\Delta S^o_{surr} > -\Delta S^o_{syst}$
$\Delta S^o_{surr} > -(-79 \ J/mol \cdot K)$
$\Delta S^o_{surr} > +79 \ J/mol \cdot K$

That was an example of the entropy change of a phase change. Solids and liquids don't change much in entropy, no matter what you do to them. However, because of the space between gas molecules, they are expandable and compressible, and can change a lot in entropy.

There are two ways the process can be done. If pressure is kept constant, while the size of the container is changed:

$\Delta S = nRln(T_{final}/T_{initial}) = nC_pln(V_{final}/V_{initial})$
n is the moles of gas
R is the gas constant, 8.31 J/mol·K
C_p is its molar heat capacity at constant pressure, in J/mol·K

If volume is kept constant, while the gas molecules are forced closer together or allowed to move further apart, by changing pressure:

$\Delta S = nRln(T_{final}/T_{initial}) = nC_vln(P_{initial}/P_{final})$
C_v is molar heat capacity at constant volume

Molar heat capacity is the amount of heat it takes to make 1 mol of a substance heat up or cool down by 1 °C. The higher the heat capacity, the more insulating a substance it is; the lower, the more thermally conducting.

Solids and liquids all have different heat capacities. Water is an example of a liquid with a high heat capacity. Compared to this, cooking oil (which really is a mixture of compounds) has a low heat capacity: it heats up faster, but by the same token cools down faster too. Meanwhile, a frying pan made of cast iron has a higher heat capacity than one made of steel.

Since solids and liquids are not very expandable or compressible, constant pressure or volume makes no difference to their C values. For gases it does, but ideally speaking all gases of the same molecular structure have the same C.

	Heat capacity at constant pressure (C_p)	Heat capacity at constant volume (C_v)
Solid	get value from book	$= C_p$
Liquid	get value from book	$= C_p$
Water	4.18 J/mol·K	4.18 J/mol·K
Gas	get value from book	$= C_p - R$
Monatomic gas	2.5R	1.5R
Diatomic gas	3.5R	2.5R

Free energy

Conveniently, with a single calculation, we can consider both enthalpy and entropy to decide if a process wants to happen. This is Josiah Gibbs' free energy (G).

$$\Delta G = \Delta H - T\Delta S$$
T is the temperature in K
ΔG and ΔH are usually in kJ/mol
ΔS is usually in J/mol·K, so has to be converted to kJ/mol·K for the calculation to work

Putting together previous observations on what processes like:

○ The most favourable situation (spontaneous forward process) occurs when ΔH is − and ΔS is +, therefore ΔG is −;
○ Also favourable: ΔH is −, ΔS is − but T is small, so ΔG is still −;
○ And: ΔH is +, ΔS is + and T is big, so ΔG is −.

Other situations are not favourable, in the sense that the forward process is not spontaneous. But, watch out, the reverse process is spontaneous!

Decide if the decomposition of water is spontaneous at 25 °C, given the following data:

	$\Delta H^o{}_f$ *(kJ/mol)*	S^o *(J/mol·K)*
$H_2O_{(l)}$	-242	110
$H_{2(g)}$	0	131
$O_{2(g)}$	0	205

$2H_2O_{(l)} \rightarrow 2H_{2(g)} + O_{2(g)}$

ΔH = products – reagents
= [2(0) + (0)] – [2(-242)] = +484 kJ/mol

ΔS = products - reagents
= [2(131) + (205)] – [2(110)] = 247 J/mol·K = 0.247 kJ/mol·K

T = 25°C + 273 = 298 K

$\Delta G = \Delta H - T\Delta S$
= (484 kJ/mol) – (298 K)(0.247 kJ/mol·K) = +410 kJ/mol
It's not spontaneous. Good thing too, since both our bodies and the earth are mostly water!

Above or below what temperature would it be?

$\Delta G = \Delta H - T\Delta S < 0$ to be spontaneous
So, $\Delta H < T\Delta S$
$\Delta H/\Delta S < T$
484 kJ/mol/0.247 kJ/mol·K < T
1960 < T
The process is spontaneous above 1960 K or 1690 °C.

Hess' Law works just as well for free energy as it does for enthalpy and entropy. In other words, **ΔG = products – reagents**.

This all works nice if your process is at STP and you can use book values. But what happens if it's not? The math gets messy. However, in intro chemistry, you can usually:

O Assume ΔH doesn't change with temperature or pressure, for a solid, liquid, or gas;
O Assume ΔS doesn't change for solids and liquids, and use the above equations to find it for a gas.

Alternately, if data are given, you can do it by using reagent and product concentrations:

$\Delta G = \Delta G^o + RTlnQ$
ΔG is the free energy change at non-standard conditions
Q is the reaction quotient

Is the decomposition of water spontaneous at 100 °C, starting with 2.7 atm H_2 and 1.1 atm O_2?

ΔG^o = 410 kJ/mol (from the last problem)
T = 100 °C + 273 = 373 K
R = 8.31 J/mol·K = 0.00831 kJ/mol·K
Q = $P^2_{H_2}P_{O_2}$ = $(2.7)^2(1.1)$ = 8.02

ΔG = 410 kJ/mol + (0.00831 kJ/mol·K)(373 K)(ln 8.02) = 416 kJ/mol
It's still not spontaneous.

In my experience, problems which use this equation often don't tell you if Q_c or Q_p is wanted, and whether Q_p should be in atm, kPa, mmHg, or Torr. Ask for clarification – or lenient grading!

Now we can finish putting thermo and equilibrium together:

ΔG is –	Forward process is spontaneous	Q < K
ΔG = 0	At equilibrium	Q = K
ΔG is +	Reverse process is spontaneous	Q > K

Given that $\Delta G = \Delta G^o + RT\ln Q$, at equilibrium $0 = \Delta G^o + RT\ln K$ or $\mathbf{\Delta G^o = -RT\ln K}$.

What is the equilibrium constant for the decomposition of water at STP?

ΔG^o = -RTlnK
410 kJ/mol·K = -(0.00831 kJ/mol·K)(273 K)(ln K)
-181 = ln K
e^{-181} = K
3.25×10^{-79} = K
This confirms that the reaction goes forward hardly at all.

The last equation can also be used for a phase change, as that is a kind of equilibrium.
To see how K changes with T, use the Van't Hoff equation:

$\mathbf{\ln(K_2/K_1) = (\Delta H/R)(1/T_1 - 1/T_2)}$

Electricity

Redox reactions, which involve a flow of electrons, can be tracked in one additional way.

$\mathbf{\Delta G^o = -nF\varepsilon^o}$
n is the moles of electrons being exchanged
F is Michael Faraday's constant, 96500 J/V, where V is volts

ε^0 is the standard potential (voltage), in V

In electrochemistry, "standard" means T is 25 °C, all gases are at 1 atm, and all solutions are at 1 M. Voltage tells how the strong the push is to move electrons from one place to another in an electrical circuit. Each half-reaction has a characteristic voltage, which is given in a table or in the problem statement. Pay attention as to whether these are given as reductions or oxidations. Then use Hess' Law to get the overall redox reaction's voltage:

○ When you flip a reaction, flip the sign on ε^0;
○ When you add reactions, add their ε^0s;
○ However, when you multiply the reaction, don't multiply ε^0.

Find the voltage of the redox reaction,
$2MnO_4^- + 8H_2O + 10Cl^- \rightarrow 2Mn^{+2} + 16OH^- + 5Cl_2$
given the standard reduction potentials:

$MnO_4^- + 5e^- + 4H_2O \rightarrow Mn^{+2} + 8OH^-$	$\varepsilon^0 = 1.51\ V$
$Cl_2 + 2e^- \rightarrow 2Cl^-$	$\varepsilon^0 = 1.36\ V$

Whoever has the higher reduction potential wants to be reduced more. So keep it the way it is and flip the other half-reaction. (Likewise, if you were given oxidation potentials, the higher one would want to be oxidized more.) Then multiply to get the same number of e⁻ in each half-reaction.

$2MnO_4^- + 10e^- + 8H_2O \rightarrow$	$2Mn^{+2} + 16OH^-$	$\varepsilon^0 = 1.51\ V$
$10Cl^- \rightarrow$	$5Cl_2 + 10e^-$	$\varepsilon^0 = -1.36\ V$
$2MnO_4^- + 8H_2O + 10Cl^- \rightarrow$	$2Mn^{+2} + 16OH^- + 5Cl_2$	$\varepsilon^0 = 0.15\ V$

At non-standard conditions:

$\Delta G = -nF\varepsilon$
$\Delta G = \Delta G^0 + RT\ln Q$:
$-nF\varepsilon = -nF\varepsilon^0 + RT\ln Q$
$nF\varepsilon^0 - RT\ln Q = nF\varepsilon$
$\boldsymbol{\varepsilon^0 - (RT/nF)\ln Q = \varepsilon}$

This is the Hermann Nernst equation. It may also be shown as $\varepsilon^0 - 2.3(RT/nF)\log Q = \varepsilon$, or $\varepsilon^0 - (0.0592/n)\log Q = \varepsilon$. The latter is what you get only when T is 25 °C, so be careful not to use it if T is anything else.

○ When ε is +, the forward process is spontaneous;
○ When it's 0, there is no net electron flow;
○ When ε is -, the reverse process is spontaneous.

Does the above reaction work at 25 °C, with all ions at 1.0 M concentration and Cl_2 gas at 0.02 M?

$Q = [Mn^{+2}]^2[OH^-]^{16}[Cl_2]^5/[MnO_4^-]^2[Cl^-]^{10}$
$= (1)^2(1)^{16}(0.02)^5/(1)^2(1)^{10} = 3.2 \times 10^{-9}$

$n = 10$ mol e$^-$

$\varepsilon = \varepsilon^0 - (RT/nF)\ln Q$
$= 0.15$ V $- (8.31$ J/mol·K$)(298$ K$)/(10$ mol e$^-)(96500$ J/V$)\ln 3.2 \times 10^{-9}$
$= 0.20$
It works!

If one of the ions involved in the reaction is H^+ or OH^-, which is often the case with redox, you can get the pH or pOH.

Which of the following ions can oxidize iron: cadmium(II), chromium(II), gold(III), or sodium? Given:
$Au^{+3} + 3e^- \rightarrow Au$ $\varepsilon^0 = +1.47$ V
$Fe^{+3} + 3e^- \rightarrow Fe$ $\varepsilon^0 = -0.04$ V
$Cd^{+2} + 2e^- \rightarrow Cd$ $\varepsilon^0 = -0.40$ V
$Cr^{+2} + 2e^- \rightarrow Cr$ $\varepsilon^0 = -0.91$ V
$Na^+ + e^- \rightarrow Na$ $\varepsilon^0 = -2.71$ V

Iron is oxidized, so flip its half-reaction:
Fe $\rightarrow Fe^{+3} + 3e^-$ $\varepsilon^0 = +0.04$ V

The other substance is reduced, so keep its half-reaction unchanged.

$Au^{+3} + 3e^- \rightarrow$	Au	$\varepsilon^0 = +1.47$ V
Fe $\rightarrow$	$Fe^{+3} + 3e^-$	$\varepsilon^0 = +0.04$ V
$Au^{+3} + Fe \rightarrow$	$Au + Fe^{+3}$	$\varepsilon^0 = +1.51$ V, which works.

$Cd^{+2} + 2e^- \rightarrow$	Cd	$\varepsilon^0 = -0.40$ V
Fe $\rightarrow$	$Fe^{+3} + 3e^-$	$\varepsilon^0 = +0.04$ V
$Cd^{+2} + Fe \rightarrow$	$Cd + Fe^{+3}$	$\varepsilon^0 = -0.36$ V, which fails.

And so on. In general, whichever reduction adds with iron's oxidation to give a positive overall ε^0 succeeds.

Gold(III) can oxidize everybody else.
Iron(III) can oxidize Cd, Cr, and Na.
Cd(II) can oxidize Cr and Na.
Cr(II) can oxidize Na.

The half-reaction potentials are defined as + or − relative to the "standard hydrogen electrode":

$$2H^+ + 2e^- \rightarrow H_2 \quad \varepsilon^0 = 0.00 \text{ V}$$

This seems to say that hydrogen's half-reaction has no voltage. But of course it is electrically active, with a voltage of about 4.5 V. Iron(III) is 0.04 V above this, so 4.54; Cr(II) is 0.91 V below, in other words 3.59. The convention exists because the hydrogen electrode is easy to make and happens to fall about halfway between the most oxidizing and most reducing substances. But it can be confusing.

Here is a picture of a complete electrochemical cell (battery):

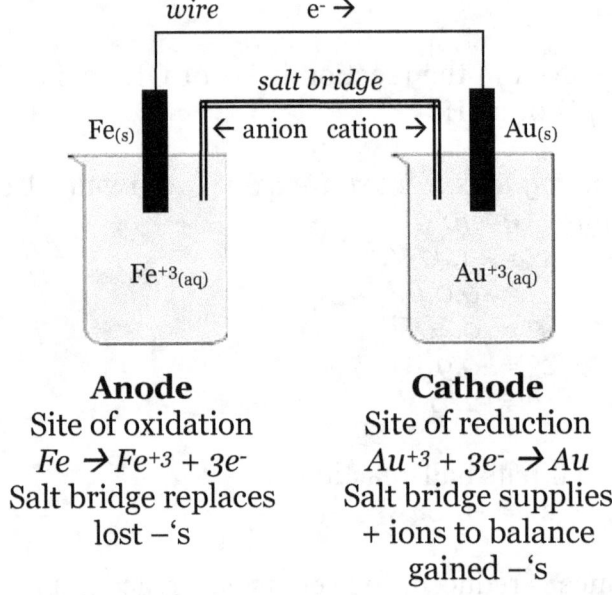

Anode	**Cathode**
Site of oxidation	Site of reduction
$Fe \rightarrow Fe^{+3} + 3e^-$	$Au^{+3} + 3e^- \rightarrow Au$
Salt bridge replaces lost −'s	Salt bridge supplies + ions to balance gained −'s

Heat

All of what we have just done is theoretical calculation. If you actually want to measure the heat, you've got to go into the lab and use a calorimeter. (It's called that because, before joules, the standard unit for measuring heat was calories.) There are two basic types:

"Coffee cup"	"Bomb"
Constant pressure (exposed to atmosphere)	Constant volume (sealed container)
Measure temperature change of contents	Measure temperature change of container and water bath

The coffee cup is made of a perfectly insulating (adiabatic) substance like styrofoam, with a loose-fitting lid and a thermometer. It doesn't change temperature or let heat in or out. This lets you assume that all of the heat from one substance or process inside it gets transferred to another.

The bomb is made of a material whose heat capacity (calorimeter constant) is known. Measure its temperature change, and you know how much heat was given off or absorbed by the substance or process inside. Typically, as shown above, you do this indirectly, by placing the calorimeter in a bath of water whose temperature you measure before and after. So, depending on how the question is worded, you may also have to take into account the amount of heat absorbed by the water bath. There has to be a way to trigger the reaction from the outside: for example, for a combustion reaction, you need to spark a fuse. The whole contraption is surrounded with insulation.

For a substance changing temperature:

$q = mC\Delta T$ or $nC\Delta T$
q is heat, in J
m is mass
n is moles
C is heat capacity, in J/g·ºC (specific heat capacity) or J/mol·ºC (molar heat capacity)
ΔT is final – initial temperature

For a bomb calorimeter:

$q = C\Delta T$
C is the calorimeter constant, in J/ºC or kJ/ºC

For a chemical or physical reaction:

q = nΔH

n is moles, usually of the main reagent of interest

What is the molar enthalpy of combustion of C_3H_8, given that burning a 3.4 g sample of it in a bomb calorimeter (C = 22.3 kJ/°C) makes the calorimeter heat up from 22.00 to 29.67 °C?

Gives heat: combustion reaction

q = nΔH

for C_3H_8, n = 3.4 g/44.11 g/mol = 0.0771 mol

q = (0.0771)(ΔH)

Takes heat: calorimeter

q = CΔT = (22.3 kJ/°C)(29.67 - 22.00 °C) = +171 kJ

Total heat = given + taken

0 = 0.0771ΔH + 171

The total is 0 because one process is exothermic (-) and one is endothermic (+).

-171 = 0.0771ΔH

-2218 = ΔH

A 70 g piece of hot iron at 287 °C is dumped in a coffee cup having 20 g of ice at - 2 °C. At what temperature do the contents settle? The heat capacities of iron and ice are 0.46 and 2.03 J/g·°C, respectively. The enthalpy of fusion of ice is 6.01 kJ/mol.

Gives heat: iron, changing temperature

q = mCΔT = (70 g)(0.46 J/g·°C)(T_{final} − 287 °C) = 32.2T_{final} − 9241

Takes heat: ice, changing temperature

q = mCΔT = (20 g)(2.03 J/g·°C)(0 − (-2) °C) = 81 J

Takes heat: ice, changing phase to water

q = nΔH

n = 20 g/18.02 g/mol = 1.11 mol

q = (1.11 mol)(6.01 kJ/mol) = 6.670 kJ = 6670 J

Takes heat: water, changing temperature

q = mCΔT

Now C is for water, 4.18 J/g·°C

q = (20 g)(4.18 J/g·°C)(T_{final} − 0 °C) = 83.6T_{final}

Total heat = given + taken

$0 = (32.2T_{final} - 9241) + 81 + 6670 + 83.6T_{final}$
$0 = 116T_{final} - 2490$
$2490 = 116T_{final}$
$21 \ ^{\circ}C = T_{final}$

It's amazing that iron so hot could heat the ice/water so little. That's because the ice and water have higher heat capacities than iron, and melting ice takes a lot of energy.

Work

Something that gives off heat can not only make another thing take it, it can make that thing do work. For example, you could burn some gasoline in an open container and just let it heat up the air nearby, or you could burn it in a closed compartment and let the pressure built up by the exhaust gases move the parts of an engine.

The sign convention for work is similar to that for heat.

Property	Sign	Meaning
Work	+	Absorbed by (done on) the system
	−	Released (done by) the system
Heat	+	Absorbed by the system (endothermic)
	−	Released by the system (exothermic)

And, as with heat, the equations for calculating work depend on the context.

For a process involving volume change, such as a gas expansion/compression or a phase change:

$w = -P\Delta V$
V is final − initial volume, in L
P has to be in kPa, since kPa·L = J

For a chemical reaction involving creation or destruction of gas:

$w = -(\Delta n)RT$
Δn is moles of gas product − moles of gas reagent
R is 8.31 J/mol·K
T is the temperature at which reaction is begun, in K

What is the work associated with the previous combustion example?

$w = -(\Delta n)RT$

$C_3H_{8(g)}$ +	$5O_{2(g)}$ →	$3CO_{2(g)}$ +	$4H_2O_{(l)}$
0.0771 mol	5(0.0771 mol)	3(0.0771 mol)	not a gas

$\Delta n = 3(0.0771) - [0.0771 + 5(0.0771)] = -0.231$ mol

T = 22.1 °C + 273 = 295.1 K
w = -(-0.231 mol)(8.31 J/mol·K)(295.1 K) = +566 J

Internal energy

Another way of saying that heat can turn into work is, all heat-work energy must be conserved. This is the First Law of Thermodynamics.

Δe = q + w
Δe is the change in internal energy, in J or kJ

Many books and teachers show this as $\Delta E = q + w$, which technically is wrong. By convention, in thermo, some variables can have an upper case (intensive, per mole) version and a lower case (extensive) one, and if so you can't mix them. For example, Δe, q, w, and Δh are in J or kJ, while ΔE, Q, W, and ΔH are in J/mol or kJ/mol. If you look carefully, when someone does a First Law calculation, they first do $\Delta e = q + w$ and then **Δe/n = ΔE.**

Find ΔE for the above combustion.

q = (0.0771 mol)(-2218 kJ/mol) = -171 kJ
w = +566 J = +0.566 kJ
Δe = (-171 kJ) + (0.566 kJ) = -170 kJ

$\Delta E = \Delta e/n$ = -170 kJ/0.0771 mol = -2210 kJ/mol

You may be asked to memorize one specific combination of ΔE, q, and w: expansion, compression, creation, or destruction of a gas by reaction or phase change. In this case, $q = n\Delta H$ and $w = -P\Delta V$, so $\Delta e = n\Delta H - P\Delta V$, or $\Delta E = \Delta H - P\Delta V/n$, which is often written as $\Delta H = \Delta E + P\Delta V$. It's not necessary to memorize this, as it follows from the equations you already know. If you do memorize it, make sure the n is there to make the intensive and extensive forms work out, and don't use the equation out of context.

Very similarly, you may be asked to memorize $\Delta S_{surr} > -q_{rev}/T$. "rev" means a reversible process, such as a phase change. In this case, as we've seen, $\Delta G = 0$, so $\Delta H - T\Delta S = 0$, $T\Delta S = \Delta H$, $\Delta S = \Delta H/T$, $\Delta S = q/nT$. That's ΔS_{syst}. Given that $\Delta S_{univ} > 0$, $\Delta S_{syst} + \Delta S_{surr} > 0$, $\Delta S_{surr} > -\Delta S_{syst}$, $\Delta S_{surr} > -q/nT$. Again, note the missing "n" and use only in an appropriate context.

In summary, there are many different ways of getting q and w (and therefore ΔE), depending on what is happening. For this reason, heat and work are called path-dependent properties. Most other thermodynamic variables are path-independent (state properties). For example, a person who boils water by heating it may have to work harder or spend more on electricity than someone who boils water by lowering the pressure, but both will create the same overall change in entropy, free energy, enthalpy, and internal energy.

Kinetics:
Can It Happen?

Overview

In thermo, we looked at the relative stability of products and reagents, only briefly mentioning that for some properties (like heat and work), the path between them matters. Kinetics is the detailed study of that path.

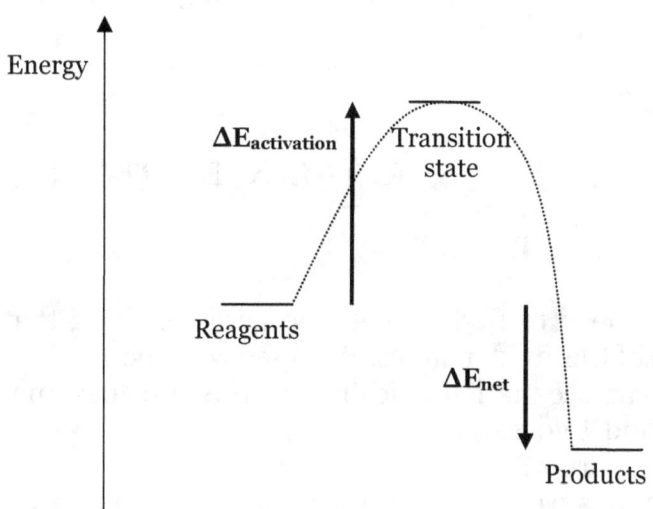

Thermodynamics	Kinetics
Products - Reagents = ΔE_{net}	Transition state - Reagents = $\Delta E_{activation}$
"What's the payoff?"	"How much work has to be done?"
Measure the amount (yield) of product	Measure the rate of product formation

The kinetic molecular theory says, in order for a reaction to succeed, reagent molecules must collide with enough energy to get scrambled. The bigger the activation energy ($\Delta E_{activation}$, or E_{act} for short), the harder they have to collide. The higher the temperature, the more energy they have when they do. The higher the concentration, the more likely collision is to happen.

A catalyst is a substance that helps bring the reagents together and/or successfully react. This lowers E_{act}. Therefore, it speeds up both the forward reaction and the reverse one, if there is one.

Differential rate

This measures how fast reaction is, depending on initial conditions.

Rate = k[reagents]^{order}
k is Arrhenius' rate constant
the order is determined experimentally

There's a separate equation for getting k.

$\ln k = \ln A - E_{act}/RT$
A is the Arrhenius constant
E_{act} is the activation energy, in J/mol
R is the gas constant, 8.31 J/mol·K
T is temperature, in K

"A" is different for different reagents, and also changes with temperature. If you run a reaction at two different temperatures...

$\ln k_1 = \ln A - E_{act}/RT_1$
$\ln k_2 = \ln A - E_{act}/RT_2$
therefore, $\ln k_1 - \ln k_2 = (\ln A - E_{act}/RT_1) - (\ln A - E_{act}/RT_2) = E_{act}/RT_2 - E_{act}/RT_1$
or:
$\ln(k_1/k_2) = (E_{act}/R)(1/T_2 - 1/T_1)$

... you get rid of the need to find A. And you get something that looks a lot like the Van't Hoff equation of thermodynamics. We'll see why soon.

At this point, you can see mathematically how the rate goes up when [reagents] goes up, E_{act} goes down, and T goes up.

Now, what about the order?

The chemical reactions we have been using up to now only show the overall change from products to reagents. On a physical level, that change may be happening in many steps. For example, for the decomposition of ozone:

O_3	$\Leftrightarrow O_2 + O$	slower
$O_3 + O$	$\Leftrightarrow 2O_2$	faster
$2O_3$	$\Leftrightarrow 3O_2$	overall

Not all steps happen at the same speed. The slowest step is the one that limits how fast the overall reaction can happen. For this reason it is called the rate-determining step (RDS). The RDS may not involve all of the reagents in the overall reaction, and it may involve some reagents (intermediates) that the overall reaction fails to show. In the above example, O is an intermediate, and the rate depends only on $[O_3]$, the reagent of the RDS.

Very often, the RDS involves 1 molecule (order) of a reagent. Sometimes, it can require 2, or 1 of one reagent and 1 of another. If you think about it, the odds of three or more molecules colliding at exactly the same time are tiny. Reagents that aren't involved in the RDS have an order of 0. And occasionally, one reagent can interfere with the others' successful collision; in that case, its order is -1.

In the reaction of A + B + C + D → E, a differential rate = k[A]¹[C]¹[D]⁻¹ = k[A][C]/[D] is found. Describe the rate-determining step.

The RDS involves a collision between A and C. B watches from the sidelines. D gets in the way.

To calculate the orders, run the reaction a few times and see how changing each reagent's concentration affects the rate.

Determine the order of each reagent in the following reaction at 25 °C:
$A + 2B + C \rightarrow 3D + 2E$

Experiment #	[A]	[B]	[C]	Initial rate
1	0.10 M	0.10 M	0.10 M	0.0043 M/hour
2	0.05	0.10	0.10	0.0011
3	0.10	0.05	0.10	0.0042
4	0.10	0.10	0.05	data lost
5	0.05	0.05	0.05	0.0021

Pick two experiments where only [A] changes: #'s 1 and 2

In general, the rate will be = $k[A]^a[B]^b[C]^c$
So, for experiment 1, $0.0043 = k[0.1]^a[0.1]^b[0.1]^c$
and for experiment 2, $0.0011 = k[0.05]^a[0.1]^b[0.1]^c$

Take a ratio to get the k's and redundant reagents to cancel
$0.0043/0.0011 = k[0.1]^a[0.1]^b[0.1]^c/k[0.05]^a[0.1]^b[0.1]^c$
$= [0.1]^a/[0.05]^a = [0.1/0.05]^a$
$3.91 = 2^a$
$\log 3.91 = \log(2^a)$
$\log 3.91 = (a)(\log 2)$
$\log 3.91/\log 2 = a$
$1.99 = a$
$2 \approx a$
The reaction is second order in A. We could have seen this qualitatively, from the fact that halving A's concentration quartered the rate.

Now look at [B].

$\text{rate}_1/\text{rate}_3 = k[0.1]^2[0.1]^b[0.1]^c/k[0.1]^2[0.05]^b[0.1]^c$
$= [0.1]^b/[0.05]^b = [0.1/0.05]^b$
$0.0043/0.0042 = 2^b$
$1.02 = 2^b$
$\log 1.02 = (b)(\log 2)$
$\log 1.02/\log 2 = b$
$0.03 = b$
$0 \approx b$

The reaction is zero order in B. We could have seen this from how changing [B] didn't change the rate.

To get [C], it would have been handy to compare experiments 1 and 4, where [A] and [B] don't change and therefore cancel when doing the math. However, since we already know their orders, we don't need them to cancel, and can take any two trials where [C] changes.

$rate_1/rate_5 = k[0.1]^2[0.1]^0[0.1]^c/k[0.05]^2[0.05]^0[0.05]^c =$
$(0.01)[0.1]^c/(0.0025)[0.05]^c = (4)[0.1/0.05]^c = (4)2^c$
$0.0043/0.0022 = (4)2^c$
$0.0043/(0.0022)(4) = 2^c$
$0.489 = 2^c$
$\log 0.489 = (c)(\log 2)$
$\log 0.489/\log 2 = c$
$-1.03 = c$
$-1 \approx c$
The reaction is -1 order in C. When [C] went down, the rate went up!

Overall, then, rate $= k[A]^2[B]^0[C]^{-1} = k[A]^2/[C]$, 2nd order in A, 0th order in B, -1 order in C, 1st order overall, since the overall order is the sum of reagent orders.

What is the value of the rate constant?

Pick any experiment, plug in the data, and solve for k.
For example, using experiment 1:
rate $= k[A]^2/[C]$
$0.0043\ M/hr = k[0.1\ M]^2[0.1\ M]$
$4.3\ 1/M^2 \cdot hr = k$
You can use data from one of the other experiments to check this value of k.

What would it be at 100 °C, if the activation energy is 6.4 kJ/mol?

$\ln k_1 - \ln k_2 = (E_{act}/R)(1/T_2 - 1/T_1)$

$k_1 = 4.3\ 1/M^2 \cdot hr$
$E_{act} = 6.4\ kJ/mol = 6400\ J/mol$
$T_1 = 25°\ C + 273 = 298\ K$
$T_2 = 100\ °C + 273 = 373\ K$

$\ln(4.3\ 1/M^2 \cdot hr) - \ln k_2 = (6400\ J/mol/8.31\ J/molK)(1/373\ K - 1/298\ K)$
$1.46 - \ln k_2 = -0.52$
$1.46 + 0.52 = \ln k_2$
$1.98 = \ln k_2$
$e^{1.98} = k_2$

$7.2 = k_2$
It makes sense that k goes up when T does.

Mechanisms

The rate of a multistep process can be approximated using the RDS alone. For example:

Step 1	A + B	→ 2C	4 hours	rate = $k_1[A][B]$
Step 2	C	→ D + E	16 years	rate = $k_2[C]$
Step 3	C + E + F	→ 2G + B	1.5 seconds	rate = $k_3[C][E][F]$
Overall	3A + F	→ D + 2G	≈ 16 years	rate ≈ $k_2[C]$

If someone asks you how long it took, you're not going to split hairs about the difference between 16 years and 16 years, 4 hours, 1.5 seconds, especially since the 16 years probably weren't measured exactly to begin with.

In the absence of experimental data, you can assume the orders are equal to the stoichimetric coefficients of the reagents.

In the above sequence, as we've seen, C and E are intermediates, substances that get produced in the course of reaction and then consumed. Meanwhile, B is an example of a catalyst: something that is there to begin with and gets regenerated in the end.

There is one catch to this. If the step before the RDS is 2-way instead of 1-way, it interacts with the RDS in a way that has to be taken into account.

Step 1	A + B	⇔ 2C	fast
Step 2	C	→ D + E	very slow
Step 3	C + E + F	→ 2G + B	very fast
Overall	3A + F	→ D + 2G	

RDS rate = $k_2[C]$

rate of 2-way step before RDS
= $k_{1f}[A][B]$ (going forward) or $k_{1r}[C]^2$ (reverse)

at equilibrium, forward rate = reverse rate, so $k_{1f}[A][B] = k_{1r}[C]^2$

Solve for [C], the intermediate between this step and the RDS
$[C] = \sqrt{(k_{1f}[A][B]/k_{1r})}$

Plug this in to the RDS rate
true RDS rate = $k_2\sqrt{(k_{1f}[A][B]/k_{1r})}$

This is the overall rate of the whole process.

The interaction between the 2-way step and the RDS is explained by Le Chatelier's Principle. In the above example, as C gets consumed in the RDS, [C] goes down and this drives the 2-way step forward, to try to replenish C.

We have just connected kinetics (little k) with equilibrium (big K):

$$K_{eq} = k_{forward}/k_{reverse}$$

And that's why the Arrhenius and Van't Hoff equations are basically the same.

Integrated rate

If you look carefully at the example of determining differential rate law orders, you'll see that we measure only initial concentrations of reagents and rates of reactions. To track how concentration and rate change over time, we need one other set of calculations. They aren't easy, so intro chemistry limits this section to reactions with only one reagent, and to the most common orders.

	0th order	1st order	2nd order
Reaction	$A \rightarrow B$	$A \rightarrow B$	$A \rightarrow B$
Differential rate law	rate = $k[A]^0 = k$	Rate = $k[A]^1 = k[A]$	rate = $k[A]^2$
Integral rate law	$[A]_t = -kt + [A]_o$	$\ln[A]_t = -kt + \ln[A]_o$	$1/[A]_t = +kt + 1/[A]_o$
Plot giving a straight line	[A] versus t slope = -k	ln[A] versus t slope = -k	1/[A] versus t slope = k
Half-time	$t_{1/2} = [A]_o/2k$	$t_{1/2} = \ln2/k$	$t_{1/2} = 1/k[A]_o$

$[A]_t$ is the concentration of A after time t has elapsed. $[A]_o$ is the initial concentration of A, when t = 0.

The integrated rate law is derived using calculus. Some teachers expect students to know how, some don't. It depends in part on how much calculus they are assumed to have studied. In case you need them, here are the proofs.

0th order
rate = k
"Rate" is the decrease in [A] as time progresses
$-d[A]/dt = k$
$d[A] = -k \cdot dt$
$\int_{[A]_0}^{[A]_t} d[A] = -k\left(\int_0^t dt\right)$
$[A]\Big|_{[A]_0}^{[A]_t} = -k\left(t\Big|_0^t\right)$
$[A]_t - [A]_0 = -k(t - 0)$
$[A]_t = -kt + [A]_0$

1st order
rate = k[A]
$-d[A]/dt = k[A]$
$d[A]/[A] = -k \cdot dt$
$\int_{[A]_0}^{[A]_t} \frac{d[A]}{[A]} = -k\left(\int_0^t dt\right)$
$\ln[A]\Big|_{[A]_0}^{[A]_t} = -k\left(t\Big|_0^t\right)$
$\ln[A]_t - \ln[A]_0 = -k(t - 0)$
$\ln[A]_t = -kt + \ln[A]_0$

2nd order
rate = k[A]²
$-d[A]/dt = k[A]^2$
$d[A]/[A]^2 = -k \cdot dt$
$\int_{[A]_0}^{[A]_t} \frac{d[A]}{[A]^2} = -k\left(\int_0^t dt\right)$
$-\frac{1}{[A]}\Big|_{[A]_0}^{[A]_t} = -k\left(t\Big|_0^t\right)$
$-1/[A]_t + 1/[A]_0 = -k(t - 0)$
$1/[A]_t = +kt + 1/[A]$

The half-time ("half-life") is how long it takes for half of the initial amount of reagent to be consumed. In other words, $[A]_t = \frac{1}{2}[A]_0$. For example:

0th order
$[A]_t = -kt + [A]_0$
$\frac{1}{2}[A]_0 = -kt + [A]_0$
$\frac{1}{2}[A]_0 - [A]_0 = -kt$
$-\frac{1}{2}[A]_0 = -kt$
$-[A]_0/2 = -kt$
$[A]_0/2k = t$

There are a couple of different ways you may see each integrated rate law written. I have chosen a form which is consistent and makes it easy to handle the following, common type of question.

What is the order of reaction for the alpha-decay of radon-220, given the following data?

Time (seconds)	Mass (mg)
0	0.0370
10	0.0326
50	0.0198
200	0.0030

See if the relationship between mass and time is linear (0th order), logarithmic (1st order), or inverse (2nd order). Here, mass is a stand-in for [A].

t	[A]	ln[A]	1/[A]
0	0.0370	-3.30	27.0
10	0.0326	-3.42	30.7
50	0.0198	-3.92	50.5
200	0.0030	-5.81	333

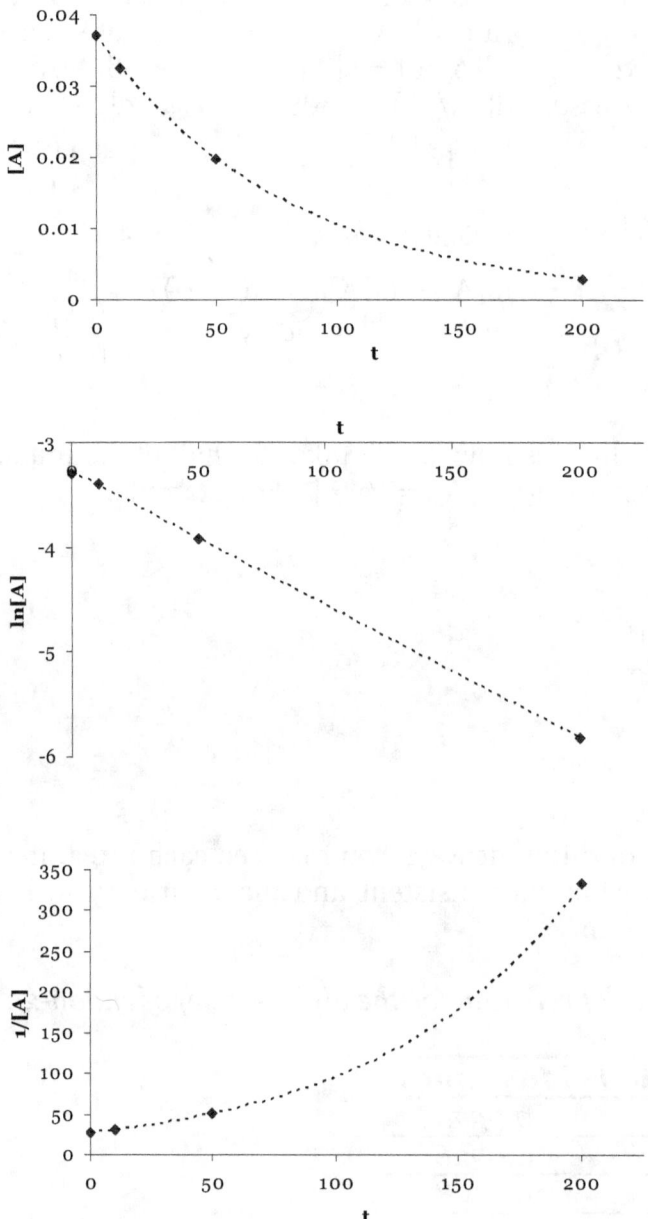

The one that gives a straight line is the right one. So, this reaction is 1st order. Most radioactive decay is.

What is the rate constant?

For 1st order, $\ln[A]_t = -kt + \ln[A]_o$
Plotting $\ln[A]_t$ versus t, the slope is $-k$
Slope = $\Delta y/\Delta x$ for any two data points, say: $((-5.81) - (-3.30))/(200 - 0$ seconds$)$
= -0.0126
This is $-k$, so k = 0.0126 1/seconds

What is the half-time?

$t_{1/2} = \ln 2/k = 0.693/0.0126 = 55.2$ seconds
Look at the data, and you'll see this makes sense: after 50 seconds, a little more than half of the starting amount was still present.

This is how the age of an object can be found, based on its content of the radioactive isotope: you know $[A]_t$ and $t_{1/2}$, so you can get k, and from there you can get $[A]_o$. For organic (carbon-based) matter – like trees, parchment, or bodies – a convenient isotope is ^{14}C, which has a half-life of 5730 years. This is "radio carbon dating".

How long will it take for your body to clear 90% of a certain poisin in air, if the initial dose is 1.3 atm and the metabolic half-life is 23 minutes?

$t_{1/2} = 1/k[A]_o$
23 min $= 1/k(1.3$ atm$)$
0.0334 1/min·atm $= k$

$1/[A]_t = +kt + 1/[A]_o$
$[A]_o = 1.3$ atm
If 90% is gone, 10% is left:
$[A]_t = (10\%)(1.3$ atm$) = 0.13$ atm
$1/0.13$ atm $= (0.0334$ 1/min·atm$)(t) + 1/1.3$ atm
207 min $= t$
This makes sense, since we are way past the half-time.

How long will it take to be 100% done?

It will never be done. After 23 minutes, 50% remains. After another 23 minutes, half of that (25%) is left. Wait another 23 minutes, and 12.5% is left. And so on. That's how half-time works, and why we use it instead of "full-time"!

Wanna try it? Stand a few metres away from a wall, facing it, with nothing blocking your way toward it. Now walk halfway toward the wall. Pause. Walk half of what's left. Pause. Keep going. Eventually, your nose will touch. But it'll take you many "half-lives" to get there. And if you were paper-thin, you would never make it.

On the other hand, you have just made it to the end of general chemistry, something it takes many people a whole schoolyear and hundreds of tuition dollars to do. Congratulations!

Index

Holy Holmium! Complete General Chemistry in 150 Pages

Summary of Formulas and Constants

Atoms

Energy of an electron in n^{th} orbit
$$E = -R^*Z^2(1/n^2)$$
E is energy in J (equal to $kg \cdot m^2/s^2$)
R^* is 2.178×10^{-18} J
Z is the number of protons

Energy of electron jump between two orbits
$$E = R^*Z^2(1/n^2_{final} - 1/n^2_{initial})$$
n_{final} of 1 is Lyman (UV), 2 is Balmer (visible), 3 is Paschen (IR), 4 is Brackett (IR)
$$1/\lambda = RZ^2(1/n^2_{initial} - 1/n^2_{final})$$
$1/\lambda$ is wavenumber, in $1/m$
R is Rydberg's constant, 1.097×10^7 $1/m$

$$E = h\nu$$
h is Planck's constant, 6.626×10^{-34} J·s
ν is frequency in $1/s$ (equal to Hz)

$$\nu = c/\lambda$$
c is the speed of light, 2.998×10^8 m/s
λ is wavelength in m

de Broglie wavelength
$$\lambda = h/mv$$
m is electron mass, 9.109×10^{-31} kg
v is velocity in m/s

Quantum numbers
n is **principal (shell)**, can = 1, 2, 3, ... ∞
l is **azimuthal (subshell)**, can = 0 (s), 1 (p), 2 (d), 3 (f), ... n-1
m_l is **magnetic (orbital)**, can = -l to +l
m_s is **spin (electron)**, can = $-\frac{1}{2}$ or $+\frac{1}{2}$

Physical properties
Pressure: 1 atm = 101.3 kPa = 760 mmHg = 760 Torr
Temperature: K = $^{\circ}$C + 273.15; $^{\circ}$F = (1.8)($^{\circ}$C) + 32
Volume: 1 L = 1000 mL = 1000 cm^3 = 1000 cc

$$n = m/M_M$$
n is moles
m is mass, in g
M_M is molar mass, in g/mol

$$n = \text{atoms of an element (or compound)}/N_A$$
N_A is Avogadro's number, 6.022×10^{23} atoms (or molecule)/mol

$$m = \rho V$$
ρ is density, in g/mL
V is volume, in millilitres (mL)

$\rho_{H2O} = 1\,g/mL$ at 4 °C

Ideal gas law
n = PV/RT
P is pressure
V is in L
R is the gas constant, 0.0821 L·atm/mol·K, 8.31 L·kPa/mol·K, 62.4 L·mmHg/mol·K, 62.4 L·Torr/mol·K, or 8.31 J/mol·K
T is temperature, in K

Nuclear binding energy (Einstein's equation)
E = (Δm)c²
E is energy in J (equal to kg·m²/s²)
Δm is lost mass, in kg
c is the speed of light

Nuclear reactions

Alpha decay (high Z): $^{b}_{a}Q \rightarrow ^{b-4}_{a-2}R + ^{4}_{2}\alpha^{+2} + 2\,e^{-}$

Beta negative decay (medium/high Z): $^{b}_{a}Q \rightarrow ^{b}_{a+1}R + ^{0}_{-1}\beta + $ antineutrino

Beta positive decay (low Z): $^{b}_{a}Q \rightarrow ^{b}_{a-1}R + ^{0}_{1}\beta + $ neutrino

Electron capture (medium/high Z): $^{b}_{a}Q + ^{0}_{-1}\beta \rightarrow ^{b}_{a-1}R + $ X-rays

Gamma decay (high Z): $^{b}_{a}Q \rightarrow ^{b}_{a}Q + \gamma$

Molecules

Polyatomic ions

CO_3^{-2} carbonate HCO_3^{-} hydrogen carbonate
$CH_3CO_2^{-}$ acetate $C_2O_4^{-2}$ oxalate
NH_4^{+} ammonium NO_2^{-} nitrite
NO_3^{-} nitrate O_2^{-2} peroxide
O_2^{-} superoxide OH^{-} hydroxide
PO_3^{-3} phosphite PO_4^{-3} phosphate
HPO_4^{-2} hydrogen phosphate $H_2PO_4^{-}$ dihydrogen phosphate
SO_3^{-2} sulfite SO_4^{-2} sulfate
HSO_4^{-} hydrogen sulfate ClO^{-} hypochlorite
ClO_2^{-} chlorite ClO_3^{-} chlorate
ClO_4^{-} perchlorate Hg_2^{+2} mercury(I)
MnO_4^{-} permanganate

Ionic lattice energy
U = MN$_A$kq$^+$q$^-$/r
U is lattice energy in J/mol
M is the Madelung constant for the type of unit cell given
N_A is 6.022×10^{23} molecules/mol
k is 8.99×10^9 J·m/C²·molecule
q^+ is the cation charge times 1.61×10^{-19} C
q^- is the anion charge times 1.61×10^{-19} C
r is the distance between the centres of + and − ions, in m

Packing efficiency = (total volume of ions/volume of unit cell)(100%)

Solubility rules

Compounds with **column I** (Na^+, K^+, etc.) ions always dissolve

NH_4^+, NO_3^-, ClO_3^-, ClO_4^-, and $CH_3CO_2^-$ always dissolve

Column VII (Cl^-, Br^-, etc.) ions dissolve, except when paired with Ag^+, Pb^{+2}, or Hg_2^{+2}

SO_4^{-2} dissolves, unless paired with column II (Ca^{+2}, Sr^{+2}, etc.) or Pb^{+2}

Anything else is only slightly soluble

Formal charge

FC = valence e⁻ - bonds − unbonded e⁻

VSEPR orbital geometries and hybridizations

Coordination number (CN) = atoms attached + lone pairs = sigma bonds + lone pairs

CN	Geometry	Angles	Hybrid (σ, LP)	Left over (π)
1	linear	-	s	-
2	linear	180	sp	p, p
3	trigonal planar	120	sp²	p
4	tetrahedral	109	sp³	-
5	trigonal bipyramidal	90, 120	sp³d	d, d, d, d
6	octahedral	90	sp³d²	d, d, d

Bond order = (bonding e⁻ − antibonding e⁻)/2

Physical processes

Standard states of the elements

Gas: He, Ne, Ar, Kr, Xe, Rn, H_2, N_2, O_2, F_2, Cl_2

Liquid: Ga, Hg, Br_2

Solid: all others

$$m_\% = (m/m_T)(100\%)$$

$m_\%$ is the mass percent of a compound in a mixture

m is its mass

m_T is the total mass of the mixture

$$m_f = m/m_T$$

m_f is the mass fraction

$$n_\% = (n/n_T)(100\%)$$

$n_\%$ is the mole percent of a compound in a mixture

n is its moles

n_T is the total moles

$$X = n/n_T$$

X is the mole fraction

$$M = n_{solute}/V_{solution}$$

M is molarity (or concentration), in mol/L

$$m = n_{solute}/m_{solvent}$$

m is molality, in mol/kg

$\Delta B_p = ik_B m$
ΔB_p is boiling point elevation, in ^oC
i is the number of particles per molecule of solute (Van't Hoff factor)
k_B is the solvent's B_p elevation constant, in $^oC \cdot kg/mol$
m is molality

$\Delta F_p = ik_F m$
ΔF_p is the freezing point depression, in ^oC
k_F is the solvent's F_p depression constant, in $^oC \cdot kg/mol$

Raoult's Law
$P = P^*_{solvent} X_{solvent}$
$P^*_{solvent}$ is the solvent's pure vapour pressure
$P_{total} = P^*_A X_A + P^*_B X_B + P^*_C X_C + ...$ (for a mixture of volatile liquids A, B, C, ...)

$\pi = iMRT$
π is osmotic pressure
M is molarity
R is the gas constant
T is temperature, in K

$KE = 3/2\ nRT$
KE is kinetic energy, in J/mol
n is moles
R is the gas constant, 8.31 J/mol·K
T is temperature, in K

Graham's Law of effusion
$v = \sqrt{(3RT/M_M)}$
v is the speed of gas, in m/s
R is 8.31 J/mol·K = 8.31 kg·m²/s²·mol·K
M_M is molar mass, in kg/mol
$v_A/v_B = \sqrt{(T_A \cdot M_{M(B)}/T_B \cdot M_{M(A)})}$ (when comparing two gases)

Boyle's Law
$P_{before}V_{before} = P_{after}V_{after}$

Charles' Law
$V_{before}/T_{before} = V_{after}/T_{after}$

Gay-Lussac's Law
$P_{before}/T_{before} = P_{after}/T_{after}$

Dalton's Law of partial pressures
$P_{total} = P_A + P_B + P_C + ...$ (for a mixture of gases A, B, C, ...)

Henry's Law of gas solubility in a liquid
$s = kP$
s is solubility in mol/L
k is the gas' Henry's law constant
P is pressure

Chemical processes

Reaction patterns
Synthesis: $A + B \rightarrow AB$
Decomposition: $AB \rightarrow A + B$
Single displacement: $A + BC \rightarrow AC + B$
Double displacement: $AB + CD \rightarrow AD + CB$

Percent yield = (actual/theoretical)(100%)

Equilibrium

Equilibrium constant
K_c = **[products]/[reagents]** (for gas and aqueous only)
K_p = $P_{products}/P_{reagents}$ (for gas only)
Each compound's concentration or pressure is measured at equilibrium and raised to the power of its stoichimetric balancing coefficient

Reaction quotient
Q_c = **[products]/[reagents]**
Q_p = $P_{products}/P_{reagents}$
Each compound's concentration or pressure is measured at any moment

$K_p = K_c(RT)^{\Delta n}$
Δn = total coefficients of gas products – total of gas reagents

$pH = -log[H^+]$
$pOH = -log[OH^-]$

$pH + pOH = 14$ for any solution

$K_w = 1.0 \times 10^{-14}$

$K_a K_b = K_w$ for a weak acid or base and its conjugate base or acid

Percent dissociation = (amount dissolved/initial amount)(100%)

Henderson-Hasselbach equations for buffers
$pH = pK_a + log([\text{conjugate base}]/[\text{acid}])$
$pOH = pK_b + log([\text{conjugate acid}]/[\text{base}])$

Thermodynamics

Hess' Law
$\Delta H_{reaction}$ = **bonds broken** (reagents) – **bonds formed** (products)
$\Delta H_{reaction} = \Delta H_{products} - \Delta H_{reagents}$
$\Delta S_{reaction} = S_{products} - S_{reagents}$
$\Delta G_{reaction} = \Delta G_{products} - \Delta G_{reagents}$
ε_{redox} = $\varepsilon_{\text{reduction half-reaction}}$ – $\varepsilon_{\text{oxidation half-reaction}}$
When you flip a reaction, flip the sign on its ΔG, ΔH, ΔS, and ε^0, invert K
When you add reactions, add their ΔG's, ΔH's, ΔS's or ε^0s, multiply their K's
When you multiply a reaction, multiply its ΔG, ΔH, or ΔS but not its ε^0, raise K to that power

2nd Law of Thermodynamics
$\Delta S_{univ} = \Delta S_{syst} + \Delta S_{surr} > 0$

3rd Law of Thermodynamics
$S_{gas} > S_{liquid} > S_{solid} > S_{solid\ at\ 0\ K} = 0$

Entropy change of a gas expansion or compression
$\Delta S = nC_p\ln(V_{final}/V_{initial})$ or $nC_v\ln(P_{initial}/P_{final})$ or $nR\ln(T_{final}/T_{initial})$
n is moles of gas
C_p is molar heat capacity at constant pressure
C_v is molar heat capacity at constant volume
R is 8.31 J/mol·K

Heat capacity of gas
Monatomic: $C_v = 1.5R$, $C_p = 2.5\ R$
Diatomic: $C_v = 2.5R$, $C_p = 3.5\ R$
Any: $C_v = C_p - R$

$C_{H2O(l)} = 4.18$ **J/g·°C**

$\Delta G = \Delta H - T\Delta S$
T is the temperature in K
ΔG and ΔH are usually in kJ/mol
ΔS is usually in J/mol·K, so convert it to kJ/mol·K

$\Delta G = \Delta G^o + RT\ln Q$
ΔG is the free energy change at non-standard conditions
ΔG^o is the free energy change at standard conditions
Q is the reaction quotient

$\Delta G^o = -RT\ln K$
K is the equilibrium constant

Van't Hoff equation
$\ln(K_2/K_1) = (\Delta H/R)(1/T_1 - 1/T_2)$

$\Delta G^o = -nF\varepsilon^o$
n is the moles of electrons being exchanged in a redox reaction
F is Faraday's constant, 96500 J/V
ε^o is the standard potential (voltage), in V

Nernst equation
$\varepsilon = \varepsilon^o - (RT/nF)\ln Q$

Spontaneous forward process: Q > K ΔG is − ε^o is +
At equilibrium: Q = K ΔG = 0 ε^o = 0
Spontaneous forward process: Q < K ΔG is + ε^o is −

Heat of a substance changing temperature
q = mCΔT or nCΔT
q is heat, in J
m is mass
n is moles
C is heat capacity, in J/g·°C (specific heat capacity) or J/mol·°C (molar heat capacity)
ΔT is final − initial temperature

Heat of a bomb calorimeter
q = CΔT
C is the calorimeter constant, in J/°C or kJ/°C

Heat of a chemical or physical reaction
q = nΔH
n is moles, usually of the main reagent of interest

Work of a process involving volume change
w = -PΔV
V is final – initial volume, in L
P must be in kPa (since kPa·L = J)

Work of a chemical reaction
w = -(Δn)RT
Δn is moles of gas product – moles of gas reagent
T is the temperature at which reaction is begun, in K

1st Law of Thermodynamics
Δe = q + w
Δe is the change in internal energy, in J or kJ

ΔE = Δe/n

Kinetics

Differential rate law
Rate = k[reagents]^{order}
k is Arrhenius' rate constant
order is determined experimentally

Arrhenius equation
lnk = lnA – E_{act}/RT
ln(k_1/k_2) = (E_{act}/R)(1/T_2 – 1/T_1)
E_{act} is activation energy, in J/mol
R is 8.31 J/mol·K

Link between equilibrium (big K) and kinetics (little k)
K_{eq} = $k_{forward}$/$k_{reverse}$

Integrated rate law
0th order: $[A]_t$ = -kt + $[A]_o$
1st order: ln$[A]_t$ = -kt + ln$[A]_o$
2nd order: 1/$[A]_t$ = +kt + 1/$[A]_o$
t is the time elapsed
$[A]_t$ is the concentration of A then
$[A]_o$ is the initial concentration of A (when t = 0)

Half-time
0th order: $t_{1/2}$ = $[A]_o$/2k
1st order: $t_{1/2}$ = ln2/k
2nd order: $t_{1/2}$ = 1/k$[A]_o$

Periodic Table of Element Numbers, Symbols, Names, and Masses

Legend: ▨ s block · ▩ p block · ☐ d block · f block

Masses in italics are for the most stable isotope of a radioactive element

1	2	3	4	5	6	7	8	9	10	11	12	13	14	15	16	17	18
1 H hydrogen 1.01																	2 He helium 4.00
3 Li lithium 6.94	4 Be beryllium 9.01											5 B boron 10.81	6 C carbon 12.01	7 N nitrogen 14.01	8 O oxygen 16.00	9 F fluorine 19.00	10 Ne neon 20.18
11 Na sodium 22.99	12 Mg magnesium 24.30											13 Al aluminium 26.98	14 Si silicon 28.09	15 P phosphorus 30.97	16 S sulfur 32.06	17 Cl chlorine 35.45	18 Ar argon 39.95
19 K potassium 39.10	20 Ca calcium 40.08	21 Sc scandium 44.96	22 Ti titanium 47.88	23 V vanadium 50.94	24 Cr chromium 52.00	25 Mn manganese 54.94	26 Fe iron 55.85	27 Co cobalt 58.93	28 Ni nickel 58.69	29 Cu copper 63.55	30 Zn zinc 65.39	31 Ga gallium 69.72	32 Ge germanium 72.61	33 As arsenic 74.92	34 Se selenium 78.96	35 Br bromine 79.90	36 Kr krypton 83.80
37 Rb rubidium 85.47	38 Sr strontium 87.62	39 Y yttrium 88.91	40 Zr zirconium 91.22	41 Nb niobium 92.91	42 Mo molybdenum 95.94	43 Tc technetium *97.91*	44 Ru ruthenium 101.07	45 Rh rhodium 102.91	46 Pd palladium 106.42	47 Ag silver 107.87	48 Cd cadmium 112.41	49 In indium 114.82	50 Sn tin 118.71	51 Sb antimony 121.76	52 Te tellurium 127.60	53 I iodine 126.90	54 Xe xenon 131.29
55 Cs cesium 132.91	56 Ba barium 137.33	71 Lu lutetium 174.97	72 Hf hafnium 178.49	73 Ta tantalum 180.95	74 W tungsten 183.84	75 Re rhenium 186.21	76 Os osmium 190.23	77 Ir iridium 192.22	78 Pt platinum 195.08	79 Au gold 196.97	80 Hg mercury 200.59	81 Tl thallium 204.38	82 Pb lead 207.20	83 Bi bismuth 208.98	84 Po polonium *208.98*	85 At astatine *209.99*	86 Rn radon *222.02*
87 Fr francium *223.02*	88 Ra radium *226.03*	103 Lr lawrencium *262.11*	104 Rf rutherfordium *261.11*	105 Db dubnium *262.11*	106 Sg seaborgium *263.12*	107 Bh bohrium *262.12*	108 Hs hassium *265.13*	109 Mt meitnerium *266.14*	110 Ds darmstadtium *271*	111 Rg roentgenium *272*							

f-block:

57 La lanthanum 138.91	58 Ce cerium 140.12	59 Pr praseodimium 140.91	60 Nd neodymium 144.24	61 Pm promethium *144.91*	62 Sm samarium 150.36	63 Eu europium 151.96	64 Gd gadolinium 157.25	65 Tb terbium 158.93	66 Dy dysprosium 162.50	67 Ho holmium 164.93	68 Er erbium 167.26	69 Tm thulium 168.93	70 Yb ytterbium 173.04
89 Ac actinium *227.03*	90 Th thorium 232.04	91 Pa protactinium 231.04	92 U uranium 238.03	93 Np neptunium *237.05*	94 Pu plutonium *244.06*	95 Am americium *243.06*	96 Cm curium *247.07*	97 Bk berkelium *247.07*	98 Cf californium *251.08*	99 Es einsteinium *251.08*	100 Fm fermium *257.10*	101 Md mendelevium *258.10*	102 No nobelium *259.10*

Feedback and Rebate Form

✂

By buying and using my book, you have already been a part of my life and career. For those of you who make the extra effort of writing to share your impressions, I offer a 10% rebate on any future book of mine. Thank you.

To participate, simply print, fill out, scan, and send as an e-mail attachment to:

Adam Gottlieb
adam.gottlieb.mail@gmail.com

Please provide your basic information.
Name →
Mailing address →
E-mail →
Phone number →
Age →
Current school level
- ○ High   Year →
- ○ College   Year →
- ○ University Year →
- ○ Other →

Job title →

How did you hear about *Holy Holmium*?
- ○ Internet search
- ○ Browsing at a bookstore
- ○ A friend recommended it
- ○ A teacher recommended it
- ○ This other person recommended it →
- ○ From the author
- ○ In the media
  - ○ TV   Show name →   Approximate date →
  - ○ Radio   Show →   Date →
  - ○ Newspaper Name →   Date →
  - ○ Magazine   Name →   Date →
  - ○ Other →   Name →   Date →
- ○ Other →

Where did you purchase it?
- ○ Lulu.com
- ○ An on-line store   Name →
- ○ A bookstore   Name →   City or town →
- ○ Other →

Who did you buy it for?
- O Myself
- O My child
- O Someone else's child
- O Mom or Dad
- O Brother or Sister
- O Friend
- O Boyfriend, Girlfriend, or Spouse
- O Coworker
- O Employee
- O Boss
- O A student of mine
- O A teacher of mine
- O Other →

Why did you purchase it?
- O To help with a class I am taking (or the person I bought it for is taking)
- O To review or learn material required for future studies
- O I need it for my job
- O It helps qualify me for a job or promotion
- O I always wanted to understand chemistry
- O To prove that I could learn or enjoy chemistry, despite previous bad experiences
- O To get someone interested in science
- O Other →

What convinced you it would fill that need?
- O The recommendation I got from someone I know
- O An on-line (internet) review
- O The title
- O The table of contents
- O Reading a few pages
- O I know the author's teaching work
- O I know his writing
- O Other →

Which chapters have you used?
- O Atoms
- O Molecules
- O Physical Processes
- O Chemical Processes
- O Equilibrium
- O Thermodynamics
- O Kinetics

Which features do you find most useful?
- ○ The writing style
- ○ The quality of explanation
- ○ The illustrations
- ○ The layout
- ○ The sequence of topics
- ○ How each topic is developed step by step
- ○ The sample problems
- ○ Summary charts and formula sheets
- ○ The index
- ○ Other →

Which do you find the least?
- ○ These ones →

Which are poorly done?
- ○ These ones →

What is the book missing?
- ○ More detail on these topics →

- ○ These whole topics →

- ○ More solved problems
- ○ Unsolved problems, with short answers at the back
- ○ An end-of-book quiz
- ○ Referral to other books and resources
- ○ Other →

Are you satisfied so far?
- ○ Yes
- ○ No

What will you do with the book now?
- ○ Keep it, because I haven't finished it
- ○ Keep it, because I may need it again
- ○ Lend it to someone who needs it
- ○ Give it to someone who needs it
- ○ Sell it
- ○ Throw it out
- ○ Other →

Will you look for other books by this author?
- ○ Yes, in chemistry
- ○ Yes, any
- ○ I don't know
- ○ No

How does this book compare with chemistry textbooks you've seen?
- ○ Better
- ○ About the same
- ○ Worse
- ○ Can't compare

How does it compare with other course notes?
- ○ Better
- ○ About the same
- ○ Worse
- ○ Can't compare

Do you want your money back?
- ○ No
- ○ Yes, because the following promise wasn't fulfilled:
    - ○ I didn't gain much knowledge
    - ○ I don't understand the material much better
    - ○ My problem-solving didn't improve much
    - ○ I didn't enjoy the book much

If you answered yes, include your receipt with this reply.

What is the highlight of this experience?
- ○ My grades improved
- ○ I qualified for a future course, degree program, or job
- ○ I understand chemistry better
- ○ I learned some interesting or valuable facts
- ○ I feel more confident about my ability to learn
- ○ I had fun
- ○ Other →

Is there anything else you'd like to say?
- ○ No
- ○ Yes →

Thank you. Best wishes to you and yours!

www.ingramcontent.com/pod-product-compliance
Lightning Source LLC
Chambersburg PA
CBHW081124170526
45165CB00008B/2536